ENVIRONMENTAL ETHICS and SUSTAINABILITY

A Casebook for Environmental Professionals

ENVIRONMENTAL ETHICS and SUSTAINABILITY

A Casebook for Environmental Professionals

Hal Taback • Ram Ramanan

CRC Press
Taylor & Francis Group
Boca Raton London New York

CRC Press is an imprint of the
Taylor & Francis Group, an **informa** business

CRC Press
Taylor & Francis Group
6000 Broken Sound Parkway NW, Suite 300
Boca Raton, FL 33487-2742

Printed on acid-free paper
Version Date: 20130619

International Standard Book Number-13: 978-1-4665-8420-4 (Hardback)

Visit the Taylor & Francis Web site at
http://www.taylorandfrancis.com

and the CRC Press Web site at
http://www.crcpress.com

To

Joyce

Hal Taback

To

Charu, Bala, Raj, and Jana

Ram Ramanan

Contents

Preface xiii
Acknowledgments xvii
About the Authors xix

Section I Environmental Ethics—Genesis and Challenges

1. Natural Instincts and Learned Behavior 3
1.1 The (Un)ethical Culture 5
1.1.1 Women of Valor and Whistleblowing 7
1.1.2 Short-Term Focus—Shareholder Value Devastation 8
1.2 Making Ethical Decisions—Six Pillars of Character 10
1.2.1 Definition of Ethics 11
1.2.1.1 Morals 11
1.2.1.2 Ethical and Ethically Neutral Values 12
1.2.2 Making Ethical Decisions 12
1.2.2.1 Why Act Ethically When It's Not in Our Self-Interest? 14
1.2.2.2 How to Recognize if a Situation Involves an Ethical Dilemma 16
1.2.3 Six Pillars of Character 17
1.2.3.1 Trustworthiness 17
1.2.3.2 Respect 19
1.2.3.3 Responsibility 20
1.2.3.4 Justice and Fairness 21
1.2.3.5 Caring 22
1.2.3.6 Civic Virtue and Citizenship 22
1.3 Ethics, Philosophy, and Religion 23
1.3.1 Religious Values 24
1.3.2 Christianity 24
1.3.3 Judaism 25
1.3.4 Hinduism 26
1.3.5 Buddhism 28
1.3.6 Islam 29
1.3.7 Enlightened Self-Interest, Agnosticism, and Atheism 31
Endnotes 34

2. Environmental Ethics and Public Policy 37
2.1 Human Beings and the Environment 37
2.1.1 Anthropocentrism 38
2.1.2 Biocentrism 39
2.1.3 Nature 40
2.1.4 Balancing Human Need and the Planet 41
2.2 Environmental Economics 42
2.2.1 Economic Theories and Public Goods 42
2.2.1.1 Myth of Markets 42
2.2.1.2 Market Externality 43
2.2.1.3 Classification of Goods 44
2.2.1.4 Tragedy of the Commons 45
2.2.1.5 Property Rights to Overcome Externality 45
2.2.2 Economic Instruments and Regulatory Policy 46
2.2.2.1 Valuation and Cost–Benefit Analysis 47
2.2.2.2 Risk Management and Resource Prioritization 50
2.2.2.3 Malthusian and Solovian Hybrid 51
2.3 Environmental Advocacy and Select Mega-Issues 53
2.3.1 Advocacy and Activism 53
2.3.2 Climate Change, Renewable Energy, and Carbon Share 55
2.3.2.1 Consensus of Scientists and Global Treaties 55
2.3.2.2 Carbon Share—The Prime Environmental Ethics Dilemma 55
2.3.2.3 Government Intervention to Contain Global Warming 56
2.3.2.4 Climate Change, Power Generation, and Renewable Energy 57
2.3.3 Hazardous Wastes and Toxic Chemicals in Products 58
2.3.4 Biodiversity—Aspect of Bioethics 60
2.3.5 Water Stress 62
Endnotes 63

3. Environmental Ethics and Corporate Governance 67
3.1 Evolution of Corporate Social Responsibility (CSR) 67
3.1.1 Advent of Corporate Social Responsibility 68
3.1.2 Morphing from Basic to Strategic Corporate Social Responsibility 70
3.1.3 Serving and Seeking Fortune at the Bottom of the Pyramid 72
3.1.4 Transforming from CSR to Creating Social Impact (CSI) 73
3.2 Framework for Reasoning to Make Ethical Decisions 75

3.2.1 Unethical Behavior—Why Does It Occur in Organizations? 75
3.2.2 Ethical Decision Making—Characteristics 76
3.2.3 Groupthink—A Dysfunctional Phenomenon 78
3.2.4 Evolution of a Framework for Making Ethical Decisions 79
3.2.4.1 Early Foray—The Four-Way Test (1932) 79
3.2.4.2 Examining the Ethics of a Business Decision (1981) 79
3.2.4.3 Ethical Values Checklist for an Environmental Professional (1998) 82
3.2.4.4 Ethical Principles for Environmental Professionals (2005) 82
3.3 Corporate Governance and Integrated Reporting 84
3.3.1 Corporate Governance and the New Social Contract 84
3.3.1.1 Recession with a Distinction 84
3.3.1.2 Concomitant Integrity Recession 84
3.3.1.3 The New Social Contract 85
3.3.1.4 Expanded Fiduciary Duty 85
3.3.1.5 Fiduciary Role—Environmental Responsibility 86
3.3.2 Integrated Financial and Sustainability Reporting 87
3.3.2.1 Environmental and Financial Factor Interrelationships 87
3.3.2.2 Environmental, Social, and Governance Performance 88
3.3.2.3 Sustainability Metrics and CSR Reports 88
3.3.2.4 International Integrated Reporting Community (IIRC) 91
3.3.3 Building an Ethical Culture 91
3.3.3.1 Corporate Governance and Global Stewardship 92
3.3.3.2 Emotional Power of Leadership Commitment 93
3.3.3.3 Legal versus Ethical Compliance 94
3.3.3.4 Warren Buffett and Salomon Brothers 94
3.3.3.5 Ethics Monitoring and Enforcement 95
3.3.3.6 Ethics Filter in Business Decisions 96
3.3.4.7 Taking Ethics Responsibility Global 97
Endnotes 97

4. Environmental Ethics Training for Leaders and Professionals 103
4.1 Building an Environmental Ethics Training Program 103
4.1.1 Evolution of Environmental Ethics Training 105
4.1.2 Evolution of Code of Environmental Ethics 105
4.1.3 Training to Make Code of (Environmental) Ethics Operative 106

4.1.4 Participatory Workshops and Case Studies 107
4.2 Conducting an Environmental Ethics Training Program 108
4.2.1 Structure of the Environmental Ethics Training Program 108
4.2.2 Ethical Decision Making—Example Case Description 110
4.2.3 Example Case Study Discussion 111
4.2.4 Values Checklist for Example Case Study 112
4.3 Educating Students and Leaders in Environmental Ethics 115
4.3.1 Educating Graduate Students in Environmental Studies 115
4.3.2 Educating Business School Students 116
4.3.3 Ethical Leadership Qualities and Mechanisms 119
Endnotes 121

Section II Environmental Ethical Dilemmas—Case Studies

5. Environmental Dilemmas—On-the-Job Conflict Situations 129
5.1 Client Attorney Directs Consultant to Not Report Findings 129
5.2 Plant Operates Substantially beyond Permit Levels 131
5.3 Boss Purposely Overestimates Client's Price for a Contract Task 133
5.4 Contractor Holds Back Completed Report for More Money 135
5.5 Expert Witness Hired to Create Reasonable Doubt 138
5.6 Colleague Seeks Help to Stop Supervisor's Suggestive Remarks 140
5.7 Plant Manager Demonstrates Compliance with Incorrect Agency Lab Data 142
5.8 New EH&S Director Has to Bring Plant under Compliance and Keep It Open 144
5.9 Agency Employee Moonlighting as Private Consultant 146
5.10 Firm Sponsors Clients on Exotic Speaking Engagements 148
5.11 Young Professional Learns the Downside of Not Blowing the Whistle 150
5.12 Supervisor Uses Company Resources for Book Not under Company Auspices 152
5.13 Production Manager Installs New Furnace in a Hurry without Proper Permit 155
5.14 Equipment Supplier Offers Incentive to Make an "Objective" Recommendation 157
5.15 Campaign Manager Advises You to Oppose Environmentally Sound Project to Win Election 159

5.16 Reporting a Toxic Solvent Leak under a Production Deadline 161
5.17 Reporting Previously Unpermitted Equipment in a Permit Renewal 163
5.18 Resampling Discharge to Override Upset Condition Sampling Result 164
5.19 Boss's Subterfuge Puts You on the Spot 165
5.20 Big Boss's Friend Expected to Get the New Consulting Contract 167
5.21 Contract Term Prevents Consultant from Revealing Problem 168
5.22 Client Believes You Overreacted 169
5.23 Your Client's Adversary Modeler Makes Error in Your Favor 171
5.24 Industry Survey Data Point Eliminated to Fit Desired Profile 172
5.25 Agency Contract Modeler Uses Excessive Conservatism 174
5.26 Consultant Debates Bait-and-Switch Option to Win Bid 175

6. Environmental Dilemmas—Judgment/Decision Situations 177
6.1 Company's Environmental Release—Where Does the Buck Stop? 177
6.2 Consulting Firm Client Fails to Report or Remediate Site Soil Contamination 179
6.3 Agency Staff Suppresses Data to Justify Unreasonable Strict Standard 182
6.4 Firm Has to Surface Competitor Deficiency without Bad-Mouthing 185
6.5 Recruiter Wants You to Bring Your Customers and Key Staff 187
6.6 Governmental Official Offers Funding Help in Return for Start-Up Partnership 193
6.7 One Firm Wants to Report Another Firm's Exceedance to Protect Public Health 195
6.8 Client Attorney Directs You to Disregard Your Additional Contrary Findings 197
6.9 Client–Attorney Privilege—Exempts Only Imminent Threat of Harm to Persons 200
6.10 Expert Has to Balance Client Confidentiality and Medical Need of the Affected 203
6.11 Firm Faces Welfare Extortion via Disingenuous Protest from Interest Group 206
6.12 Plant Manager Installs Emission Control Equipment without Proper Permit 208
6.13 Consulting Firm Bills Client at a Higher Senior Rate for Junior-Level Work 210

6.14 Firm Offers Country Club Membership for Procurement Officer ... 212
6.15 Employee Notices Dishonest Use of Company Resources by a Colleague ... 214
6.16 Hazardous Waste Is Accidentally Sent to a Municipal Landfill ... 216
6.17 Risk Communication—Numbers but No Panic ... 218
6.18 Agency Inspector Ponders—To Cause a Plant Closure or Not ... 220
6.19 TV Reporter Pressured to "Get the Dirt" on Local Industry ... 222
6.20 New Data Blows Industry's Agreement with EPA ... 223
6.21 Should Consultant Use a Bounty Hunter Provision on Client? ... 226
6.22 Environmentalist Blockade of Offshore Incineration Ship ... 227
6.23 Driving 55 mph Only When the Cop Is Around ... 229
6.24 Biased Agency Subordinate Stonewalls a Project ... 230
6.25 Permitting Engineer Softens NSR Rules to Help Local Industries ... 231

Glossary ... 233

Appendix ... 241

Index ... 243

Preface

Ethics—The Fourth Bottom Line

"Earth provides enough to satisfy every man's needs, but not every man's greed."[*] Extreme greed, whether for money or nature's resources, has disastrous consequences. Today, given the unprecedented size and speed of global transactions, the world is poised for potential tsunamis in many areas because "the avenues to express greed have grown so enormously"[†] and "man has acquired significant power to alter the nature of his world."[‡] In this book we focus on sustainability, an area we can influence, and expand the triple bottom-line context of people, planet, and profit with a fourth component—purpose. People are inherently self-serving. Absent purpose as a moderator, one could easily skew the objective through the inherent bias of self-interest—caring only for people who mimic us, protecting only parts of the ecosystem that overtly serve us, profiting only a subsection of the stakeholders, misleading investors through underreporting material environmental risks from climate change, misleading consumers through "green washing," or overclaiming the environmental friendliness of products and services.

Path Taken Matters

Organizations are run by people, and it is the individuals who are responsible for the actions of, and the consequent outcomes and impact on, the corporations and society. For instance, U.S. corporate law does not mandate maximizing share price or shareholder wealth, yet myopic focus on

* Govind Singh, "Mahatma Gandhi: A Sustainable Development Pioneer," *Eco Localizer*, http://ecowordly.com/2008/10/14/mahatma-gandhi-who-first-envisioned-the%20concept-of-sustainable-development (last modified October 14, 2008); http://www.mkgandhi.org/articles/environment1.htm.

† "Testimony of Alan Greenspan," presented during the Federal Reserve's Monetary Policy Report: Federal Reserve Board, July 16, 2002, http://en.wikipedia.org/wiki/Alan_Greenspan (accessed December 2012).

‡ Rachel Carson, *Silent Spring* (Boston: Mariner Books, 2002), http://www.goodreads.com/work/quotes/880193-silent-spring (accessed December 2012).

short-term return is driven by the skewed executive compensation system. The BP Gulf oil spill in 2010 is a clear example of this culture of cutting corners that led to the all-around loss of regional ecosystems, oil, fishing, and tourism industries, as well as the shareholders and bond holders of BP. "End does not justify the means"*—the path taken matters. The extent to which people will go to achieve their personal objectives often depends on their character, which is formed at an early age. Since the potential for cheating exists in childhood and is carried into adulthood, ethical values should be learned much earlier in life. Therefore, companies are faced with the challenge of making significant additional efforts to ensure that their employees unlearn some of this ingrained unethical behavior. What makes a person behave ethically (or unethically)? Our inherent self-preservation instinct is something we must recognize and deal with rationally whenever we encounter situations that tempt us to stray from the right course of action. Ethics training is needed to combat this natural instinct. Just as athletes must train their bodies to perform at their peak, environmental (and other) professionals require ethics training to ensure that they continue to behave ethically in difficult situations.

Making Ethical Decisions

A fundamental definition in this book is as follows: *Ethics is the difference between what a person has the right to do and the right thing to do!* A person's "right to do" is often defined by the law of the land. The "right thing to do" is the action taken in response to a situation that will result in the greatest benefit and the least harm to all the stakeholders. Ethical decision making is affected by three qualities: competence in identifying issues and evaluating consequences, self-confidence in seeking different opinions and deciding what is right, and willingness to make decisions when the issue has no clear solution. The first aspect involves the ability to distinguish right from wrong, good from bad, and propriety from impropriety. The second is self-confidence built through real-world experience and training. The third is a commitment to do what is right and proper. Ethics is an action concept, not simply something to think about and rationalize. It involves conduct and behavior. Ethics requires character and courage to meet the challenge when doing the right thing costs more than the stakeholder wants to pay.

* Anthony J. Perel, *Gandhi, Freedom, and Self-Rule* (Lanham, MD: Lexington Books, 2000), p. 75.

Environmental Ethics Training

Building a culture of ethics is critical, and effective ethics training is crucial to overcome our inherent selfishness. Universities teach ethics with hypothetical dilemmas, companies use a "code of ethics" that their employees sign, and books are written on humanity's greed and its inhumanities to animals and plants. But this book is different in that it concentrates on providing hands-on practical training. This book serves as our first, but very important, step in providing society with training tools to stem the rot and rebuild an ethics culture, one piece at a time. Frequent in-house ethics training and organizational ethical culture-building supplemented with an appropriate incentive/deterrent system will help develop ethical values and minimize the temptations to cheat. The objective of this book is to serve as a resource for training environmental practitioners and students aspiring to become environmental leaders and professionals. The suggested approach is to educate and train professionals and practitioners through participative workshops that discuss relevant, real-world dilemmas and help people learn how to do the right thing, rather than to leave it to their instincts.

Construct of This Book

Ideally, this material will be used by university professors and organizational environmental managers and leaders for educating students and training employees who are performing environmental duties for developing and building an ethical culture. Additionally, groups of environmental professionals or individuals may use it on a scheduled periodic basis to maintain individual awareness of the ethical dilemmas that they may encounter in their practice and in maintaining their professional board certifications. These participatory training workshops are designed to help our professionals hone their skills and prepare our students to develop their skills to make ethical decisions. These could serve as one significant step to counter the myriad temptations and opportunities to stray that entice all of us in today's business world. This book is not intended to outline theories of ethics and the terms used in ethical debate, but rather to examine the conflicts experienced by environmental professionals using actual dilemmas and solutions thereof. Some ethical choices are clear-cut, while others are more problematic; it is by looking into the problematic ones that we can start to appreciate the issues involved in resolving those dilemmas.

Section I of this book takes an incisive look at the genesis and challenges of environmental ethics. Chapter 1 starts with natural human instincts, the unethical culture all round, and learned behavior through religious and philosophical thought processes. Chapter 2 covers the human attitude and treatment of other species and the natural world, as well as some fundamental concepts in environmental decision making in the public policy arena, especially on select mega-environmental issues. Corporate governance and the significance of building ethical cultures in organizations in the context of the new business ambience and social contract are presented in Chapter 3. The final chapter in Section I provides insight and specifics on how to develop and conduct an environmental ethics program to educate and train leaders, professionals, and students.

Furthermore, a rich tapestry of diverse real-world dilemmas that are relevant to environmental professionals has been devised and included in Section II. It identifies and deals with numerous dilemmas in a case study format, offering various options, testing ethical values, and giving trainees practice in the challenge of doing the right thing. Additional material is available from the CRC Press website: http://www.crcpress.com/product/isbn/9781466584204.

This casebook on ethics offers to help the environmental professional appreciate the choices that others make and evaluate the justification they give for those decisions. It also sharpens one's ethical awareness; that is, it invites the reader to undergo a conscious examination of values and choices and how these may have shaped his or her life so far and how they can be used to shape it in the future.

Hal Taback
Carlsbad, California

Ram Ramanan
Chicago, Illinois

Acknowledgments

I thank my clients, colleagues, family, and staff who have supported me over the past 35 years while I have done the right thing in protecting the environment and public health while still meeting my other priorities. I thank my associates of over three decades at the Air and Waste Management Association where I have chaired its ethics committee; Bob Hall, a long-term associate and past president and executive director of A&WMA; and Lisa Bucher, the editor of *EM* magazine, where my ethics columns were published for five years. They have allowed me to identify real-world ethical issues and encouraged me to formalize ethics training in columns, papers, workshops, and now, this book.

Hal Taback

I thank my academic, corporate, and consulting peers of over four decades who have inspired me to pursue an ethical path throughout my career and who have offered me the support and provided me the courage to never stray from my dedication to strive for excellence and to do so with a clear conscience. I thank my parents and parents-in-law for instilling those values in me through exemplary living. My father's graduation gift, a plaque etched with Einstein's quotation, "The concern for man and his destiny must always be the chief interest of all technical effort. Never forget it among your diagrams and equations," drove me to pursue the environmental field. Several decades later, my children's retirement gift, another plaque, this one with Gandhi's quotation, "Earth provides enough to satisfy every man's need, but not every man's greed," opened my eyes to the fourth dimension—ethics. Today I am consumed by the desire to build our future leaders with a better comprehension of purpose and achieve a higher form of capitalism.

Finally, I want to thank my daughter's childhood friend, Alice Bootman, for her unstinting help in editing the drafts of this book.

Ram Ramanan

About the Authors

Hal Taback, PE, BCEE, QEP, REA, is president of Hal Taback Company of Carlsbad, California, a Southern California environmental consulting firm. He has over 40 years of professional engineering practice in regulatory and engineering issues related to reducing the risk from toxic air pollutants and hazardous waste and air compliance issues. For eight years he served as the principal consultant to the American Petroleum Institute, coordinating with the United States Environmental Protection Agency on developing Maximum Achievable Control Technology rules for all processes in the petroleum industry.

Taback holds a master of science in engineering from Princeton University and a bachelor of mechanical engineering from the University of Rochester. He is a licensed professional engineer (PE) in California, Hawaii, Minnesota, Oklahoma, and Arizona; an American Association of Environmental Engineers board-certified environmental engineer; a qualified environmental professional; an ANSI/RAB-certified ISO 14001 EMS auditor; a California-registered environmental assessor (REA); and a member of Sigma Xi and Tau Beta Pi.

A fellow and honorary member of the Air and Waste Management Association (A&WMA), Taback chairs its national ethics committee. He has chaired ethics workshops at eight national meetings of the Air and Waste Management Association and has lectured on the ethics of the environmental professional to technical and bar associations across North America. He has over 140 publications in the environmental field. For five years Taback wrote a bimonthly column in *EM* magazine entitled "Ethics Corner," in which he discussed solutions to common ethical dilemmas experienced by environmental professionals. His papers on ethics have been published in *Environmental Auditor*, *Engineering Times*, and *Southern California Environmental Digest*. Taback's presentations on ethical obligations of the environmental professional are conducted as an interactive discourse using ethical dilemmas as the basis for discussion.

Dr. Ram Ramanan is a Fulbright fellow and Air and Waste Management Association (A&WMA) fellow with over 40 years of corporate, consulting, and academic experience in the management of sustainability, environmental, health, and safety issues. He is the interim director of the Center for Sustainable Enterprise and an industry associate professor at the IIT Stuart School of Business. In addition, he currently chairs the Education Council and serves as director on the International Board of Air and Waste Management Association.

Dr. Ramanan, an ExxonMobil retiree, has developed and implemented corporate risk management and sustainability strategies and programs, managed corporate liability at dozens of Superfund sites, and advised several mega (>$1 billion) new ventures on environmental and social impact assessments in Asia, Africa, and the Americas. He has also led industry group advocacy efforts in environmental standards development at the state and federal levels in the United States, India, and at the World Bank. Dr. Ramanan has also served as fellow and director at ICF International, as president of a subsidiary of Trinity Consultants, and as department manager at AECOM (formerly ERT). Prior to that, he held various positions, from design engineer to general manager, providing air pollution control systems for the power and petroleum sectors in India.

Dr. Ramanan has published or presented over 40 technical and management papers throughout the world. He holds a PhD from the University of Texas at Dallas, an MBA from the University of Texas at Austin, an MS in marketing and engineering from the University of Bombay, and a BS in chemical engineering from the Indian Institute of Technology. He is an American Academy of Environmental Engineers board-certified environmental engineer, a professional engineer in Texas and India, a chartered engineer and chartered environmentalist and scientist in the UK, a qualified environmental professional, and an elected fellow of the Indian Institute of Engineers and the Indian Institute of Chemical Engineers.

Section I

Environmental Ethics—Genesis and Challenges

1

Natural Instincts and Learned Behavior

People are inherently self-serving. The extent to which people will go to achieve their personal objective often depends on their character, which is formed at an early age. From birth, a child has natural self-preservation instincts and inherently selfish behavior. A baby has an instinct to cry until it is fed and comforted by being held and fed. When denied this comfort for a brief time, babies obtain attention by crying. Later, in elementary school, children grab toys from one another saying, "Mine, mine, mine," while the teacher is saying, "Share, share, share." Children often lie to avoid punishment or to obtain something they feel would not be available to them otherwise. The degree to which children behave in this way varies and depends on what they have learned from and been trained to do by their family and other social contacts.

Other factors that influence ethical behavior in later life are youth sports, retail sales jobs, and the desire for academic achievement. Youth sports often have parent spectators that cheer for their team and verbally taunt the opponent, and through this behavior a child learns that winning is important. As that child reaches high school, his or her coach might create an impression that winning is "everything." In some sectors of society, families have a need for their children to get into the best kindergarten, the best elementary school, the best high school, and ultimately the best university.

In the environmental arena, cheating most often manifests itself in the form of misleading investors through underreporting material environmental risks from climate change,[1] misleading consumers through "green washing,"[2] or overclaiming the environmental friendliness of products and services.

While the geopolitical issues of the world seem poised for potential tsunamis in many areas, in this book we focus on sustainability, an area we could influence. However, we do expand the triple bottom-line context of people, planet, and profit with a fourth component—purpose. Absent purpose as a moderator, one could easily skew the objective through the inherent bias of self-interest—caring only for people who mimic us, protecting only parts of the ecosystem that overtly serve us, and of course profit only a subsection of the stakeholders.

In recent decades this aspect has gained critical significance—driven by the potential for creating the perfect storm that could eventually lead to a tsunami. In the profit aspect, according to Alan Greenspan, "It is not that humans have become any greedier than in generations past. It is that the avenues to express greed had grown so enormously."[3] The analogy in the people and planet aspects is, as Rachel Carson said, "Only within the moment of time represented by the present century has one species—man—acquired significant power to alter the nature of his world."[4]

There are several avenues where people learn what to value and how to behave—at home, from peers, and at religious institutions, if they belong to one. While the concept of formal ethics education and training is not new, proponents like universities, government entities, and industrial firms often take an analytical and prescriptive approach to teaching ethics. Universities teach the background and logic of ethics with hypothetical dilemmas, like two people on a raft with food for only one to survive. Who should be sacrificed? Companies may, for instance, develop a "code of ethics" that their employees are to read and acknowledge once or twice a year and sign a statement of acknowledgment. Books are written on the instincts of man to succeed and man's inhumanities to animals and plants.

But, this book is different in that it concentrates on providing hands-on practical training. The objective of this book is to serve as a resource for training environmental practitioners and students aspiring to become environmental leaders or professionals. The approach is to educate and train professionals and practitioners through participative workshops that discuss real-world relevant dilemmas and help people learn how to do the right thing rather than leave it to their instinct. Furthermore, a rich tapestry of diverse real-world dilemmas that are relevant to environmental professionals has been devised and included in Section II of the book. They could draw from one of these dilemmas that they experienced in a training session and adapt or adopt it to resolve the environmental dilemmas they face and generate an appropriate response.

What makes a person behave ethically (or unethically)? A common misconception is that people are basically good and will usually do the right thing in business dealings, which in recent times has proven to be untrue. Is there something in the human brain that prompts us to do the right thing in certain situations? Do we inherently care about other people? No. Ethical behavior is learned and humans are born with an instinct for self-preservation (which sometimes must be curbed) that makes us naturally selfish, and this instinct remains with us throughout our lives. Our inherent self-preservation instinct is something we must recognize and deal with rationally whenever we encounter situations that tempt us to stray from the right course of action. Ethics training is needed to combat this natural instinct. Just as athletes must train their bodies to perform at their peak, environmental (and other) professionals require ethics training to ensure that they continue to behave ethically in difficult situations.

1.1 The (Un)ethical Culture

The explosion of catastrophic events driven by integrity failures is mind-boggling. As discussed in Section 1.1.2, the potential consequences of unethical actions have reached monumental proportions with the world becoming flat and the unprecedented size and speed of global transactions. Here we step back and take a look at what has been happening to our society from the perspective of ethics. In his book *The Cheating Culture*, David Callahan identifies many examples of cheating, ranging from shoplifting and music theft to stock market pundits promoting declining securities to plagiarism and false citations by book authors, as well as facts manufactured by editorial writers at major newspapers regarding high-profile corporate scandals. He also discusses hot-button topics such as large automotive service providers, attorneys, and other professionals who continuously overcharge clients, corporate accounting firms that cook the books, cops who take bribes, clergy who bilk parishioners, athletes who use performance-enhancing drugs to win, and corporate executives who shower themselves with excessive incomes even in times of reduced corporate performance. Reading Callahan's book would lead one to believe that everyone is cheating.[5]

Cheating is not a recent phenomenon, however. Callahan references a 1964 study conducted by William Bowers,[6] in which three-quarters of college students surveyed admitted to cheating. This included changing grades, bribing teachers, getting parents to intimidate teachers, and stealing tests. All of this cheating was justified, according to respondents, by the pressures of getting into a good university that would eventually lead to high-paying jobs. Getting into a top school was rated above ethics by both parents and students surveyed.

This is indeed a dangerous trend. Coupled with the potential for devastating consequences in today's context (as discussed in the previous section), this trend has to be stemmed, and immediately. Given this background of employees entering the workforce, organizations are starting with a mega-challenge to build and sustain an ethical culture. Callahan opines that the U.S. government is attempting to increase enforcement and penalties to filter out corporate corruption. He suggests it is the government's responsibility to ensure the ethical practice of corporations. As an illustration, he cites the example of a government sting operation, called Operation Ill Wind, during which corrupt contractors were caught and punished for offering bribes and inflating costs. In addition, the Federal Sentencing Guidelines, which apply to the sentencing and punishment of companies convicted of environmental crimes (e.g., violations of U.S. Environmental Protection Agency regulations), now require that offending companies institute an ethics program. Callahan maintains, however, that current government and corporate ethics programs are not working. He argues that Americans lack the necessary opportunities to derive meaning from life not focused solely on making money and spending it. He proposes that every year a significant amount of taxpayer money and corporate assessments be moved to the nonprofit sector, professional societies, or other nongovernmental organizations (NGOs) for ethics training.[7]

Callahan claims that economic fairness is not feasible without a more robust government. He says that we should create more "livable" communities (i.e., bring back close-knit neighborhoods), reduce "grotesque materialism" (i.e., possibly institute a consumption tax), and develop a sense of "shared moral purpose" with regard to racial and ethnic diversity. Another suggestion Callahan offers is the reinstitution of school honor codes. He suggests teaching programs in schools that promote character education based on ethical values such as fairness, honesty, and justice. He says that young people should be taught to look beyond their own narrow self-interest, a mindset critical to strengthening the standards of academic integrity. Finally, he believes business and professional schools should teach and study ethics.

Lynn Sharp Paine, a professor at Harvard Business School, offered the following observation: "Today's leading companies are expected not only to create wealth and produce superior goods and services but also to conduct themselves as 'moral actors'—as responsible agents that carry out their business within a moral framework … thus … society today has endowed the corporation with a moral personality."[8] Professor Paine does not, however, explain how companies might accomplish this moral behavior. How can companies be held responsible for not developing in their employees a set of ethical values when the employees have been exposed to a lifetime of cheating outside the company? Since the potential for cheating exists in childhood and is carried into adulthood, ethical values should have been learned much earlier in life. Therefore, companies are faced with the challenge of making significant additional efforts to ensure that their employees unlearn some of this ingrained unethical behavior. Frequent in-house ethics training and organizational ethical culture building supplemented with an appropriate incentive/deterrent system will help develop ethical values and minimize the temptations to cheat.

Callahan suggests government intervention using taxpayer money or consumption tax to educate students to reverse this trend; Paine recommends that companies add moral behavior to other responsibilities. All of these are great ideas, but Callahan, Paine, and the other academic ethicists fail to detail just how to undertake the tasks of teaching people ethics. This book will continue where Callahan's book left off, although it will be focused more narrowly on the issues that perplex environmental practitioners.[9] This book serves as our first, but very important, step in helping society with training tools to stem the rot and rebuild an ethics culture, one piece at a time.

1.1.1 Women of Valor and Whistleblowing

When *TIME* Magazine was to announce its choice for "Person of the Year," in early 2003, many thought George W. Bush was the obvious choice due to his response to the 9/11 attacks. Instead, it selected three of the most prominent whistleblowers of all time as Person of the Year. All three of them (Sherron Watkins and Cynthia Cooper, both vice presidents at Enron and WorldCom, respectively, and Coleen Rowley, staff attorney with the FBI) had come forward in the prior two years, and their impact was incredible. As the lead for *TIME*'s article read: "They took huge professional and personal risks to blow the whistle on what went wrong at Enron, WorldCom, and the FBI—and in so doing helped remind us what American courage and American values are all about."[10]

These three women of extraordinary courage and sense for doing the right thing had an impact on the world as profound as any other person in a high government office has had. Watkins warned Enron CEO Ken Lay and Cooper the WorldCom board of improper accounting practices, and Rowley alerted FBI Director Robert Mueller about how the Minneapolis field office plea to investigate 9/11 co-conspirator Zacharias Moussaoui was ignored by the headquarters. These women did the right thing just by doing their jobs with courage and conviction while putting their professional future on line. Their jobs, their health, their privacy, their sanity—they risked all of these to bring us badly needed honesty about the trouble inside crucial institutions. Although environmental professionals may never face a financial dilemma of such proportions, it is critical that people can trust in us and in the integrity of our profession because we are dealing with human health and the environment. Following the example of these whistleblowers, ethical environmental professionals must serve as a fail-safe mechanism that does not falter in the protection of public health.

Who are these women? Where did they find the courage? Who were the stakeholders in their lives that they had to consider? Can we, as ordinary middle-class individuals with dependents and a home, ever find ourselves willing to go to such extremes to do the right thing? Because they are women, some might draw the quick conclusion that women have less at stake in their organizations and so might be more willing to expose weaknesses. However, for each of them, as primary breadwinners, the decision to confront the higher-ups meant jeopardizing the paycheck on which their families truly depended.

The *TIME* article says that studies have shown that women are actually a bit less likely than men to be whistleblowers. Whistleblowers don't have an easy time. These women initiated the cleanup of professional life in America that will have a lasting impact. It is likely that no one could have prevented the 9/11 attacks or eliminated the reoccurrence of Enron- and WorldCom-like fiascos, but it is people with jobs similar to those held by Watkins and Cooper who are the best positioned to bring such events to light. The new law that requires CEOs and CFOs to vouch for the accuracy of their company's books is just one step along in the country's corporate regulatory revolution. The *TIME* article reminds us that Watkins, Rowley, and Cooper are ordinary people who did not wait for higher authorities to do what needed to be done. One of the characters in Ibsen's play *An Enemy of the People*, about unwelcome truth telling, reminds us of what we admire about this dynamic trio: "A community is like a ship," he observes, "everyone ought to be prepared to take the helm."[11] When the time came, these women took the helm. *TIME* also quotes Martin Luther King, who said, "Our lives begin to end the day we become silent about things that matter."[12] These three women were exemplary role models who displayed the courage to do the right thing in the face of adverse consequences to themselves and their families. Organizations are run by people, and it is the individuals that are responsible for and impact the ethical outcomes of corporations and society. Building a culture of ethics is critical, and effective ethics training is crucial to overcome our inherent selfishness.

1.1.2 Short-Term Focus—Shareholder Value Devastation

The role of business is transforming from one merely fulfilling a social contract to taking on social responsibility with the growing recognition that shareholders are only one of many stakeholders. Upon the demise of Enron, at a time when increasing share price was the mantra, several scholars began to question shareholder primacy and discovered that U.S. corporate law does not mandate maximizing share price or shareholder wealth. A myopic focus on short-term return is driven more by the skewed executive compensation system, with practices like collusion to share the goodies. Perhaps a short-term focus is exacerbated by the emergence of speculators pushing derivatives rather than investors building long-term value. The following three examples cover a spectrum of unethical behavior, from BP cutting corners on required maintenance expenses to achieve next quarter returns, to Enron executives creating complicated illusory and fraudulent accounting to mislead shareholders and lenders, to Madoff's downright cheating of uninformed speculators in a Ponzi scheme of investments. The BP disaster was in the environmental and safety arena; the Enron and Madoff debacles are drawn from other spheres.

The BP Gulf oil spill in 2010 was a clear example of an all-around loss of immense proportions: regional ecosystems and the oil, fishing, and tourism industries were all big losers, but the biggest losers were the shareholders and bond holders of BP. Stout cites the events leading to the spill as the "dumbest idea in the world."[13] Investigating government authorities identified the root cause to be chronic safety lapses, which could be traced to a systemic lack of environmental ethics culture. BP and its contractors cut short-term costs and thereby ignored standard safety procedures. However, when BP CEO Tony Hayward, under whose watch the spill occurred, tried to palm off the blame on contractors, he lost his job due to public outrage. In 2012 BP was indicted by the U.S. federal grand jury for corporate manslaughter and was sentenced to a criminal penalty of $4 billion, and the two highest-ranking supervisors were also indicted on charges of involuntary manslaughter. Furthermore, BP will pay about half a billion dollars for having calculatingly lied to investors about the amount of oil spilling into the Gulf. The U.S. government has temporarily banned BP contracts to supply fuel to the U.S. Department of Defense. The EPA said it imposed the ban because the company's conduct during the 2010 *Deepwater Horizon* disaster showed a "lack of business integrity."[14] This event highlights several factors pertinent to environmental professionals and the cost of an unethical culture. The leaders and managers indulged in promoting short-term cost cutting related to safety and environmental maintenance expenses. Even the CEO indulged in lying to the regulatory authorities and in passing the blame on to contractors. BP broke nearly every one of the six pillars of character (described in Section 1.2.3) that form the basis of ethics. And, yes they paid a heavy price in fines and settlements, cost of cleanup, loss of reputation, loss of market capital, and indictment of personnel. In addition, intangibles such as employee morale and investor and lender distrust were some of the other factors.

Gandhi said, "Earth provides enough to satisfy every man's needs, but not every man's greed." Extreme greed, whether for money or nature's resources, has disastrous consequences. The Enron fiasco is a classic example of how executive greed can destroy a major corporation within a short span of time. Sims attributes culpability in the demise of Enron, a scandal revealed in early 2001, to irresponsible and unethical leadership.[15] There was an unethical leadership avalanche. The leaders were using short-term shareholder wealth creation as a veil and creating an illusion that the primary standard was doing what was best for the company, while the executives were running a risk-free company for themselves. The culture became one in which ethics issues were not addressed until they became significant, and even then were treated as optional, not obligatory! Collusion—partnership in crime—became rampant. Some employees were happy to accept stocks and bonuses for supporting unethical deals. Executives misled the public using questionable transactions off the balance sheet. Speculative investors went with the high-risk, high-reward Enron stocks that were rising asymptotically. Banks and brokerage houses, to earn fees and huge profits, invested in high-risk partnerships that Enron used to hide debts. Then Enron auditor Arthur Anderson, who also served as a consultant and made over $50 million, played along. Finally, Enron attorneys Vinson & Elkins ignored warnings of impropriety from Sherron Watkins and wrote creative special purpose partnerships. Ethical standards regarding conflict of interest were waived at the company, which allowed one senior executive to set up private partnerships and another to order shredding of documents that were under discovery. When it got too hot, each executive sought to minimize personal responsibility for ethical issues—they blamed each other or feigned ignorance of illegality. The result was that true investors looking for long-term value, as well as long-term employees, saw their investments and lifetime savings evaporate. Unethical behavior from any sphere provides a lesson. In the case of Enron, the company evaporated, not much unlike the indictment of tobacco companies for fraudulently covering up the health risks associated with smoking and for marketing their products to children. The tobacco industry and its collusion in hiding the health effects and use of nicotine to promote addiction comes to mind as the best known example of cheating in the environmental, health, and safety sphere.

Finally, the Bernard Madoff multi-billion-dollar Ponzi scheme was revealed in 2008. His success was predicated on investors receiving abnormally high and quick returns without questioning his methods. His conviction (150 years in prison) was not due to whistleblowing but was the result of complaints by jealous investors who did not get similar abnormal returns. The returns on investments promised and offered to the early investors of the Madoff Ponzi scheme were several folds larger than returns from the broad stock market. This gives credence to the likelihood that these were speculators and not serious investors who were enticed and wanted to make very high gains in the very short term. There is no ethical dilemma involved; this was pure crime. There are several parallels that could be drawn from illegal dumping or exporting of hazardous waste by companies and then getting indicted. The above discussions will serve as background lessons of potential consequences of unethical behavior as we address real-world applications of learning from these experiences to solve dilemmas that environmental professionals may experience. Leadership, culture, and training are the three major takeaways.

1.2 Making Ethical Decisions—Six Pillars of Character

Ethical decision making is the process of considering and selecting one of several alternatives to ensure that an action is ethically sound. There may be a number of possible ethical options to consider, not all equally meritorious. Making ethical choices can be complex because in many situations there are winners and losers involved, with varying levels of impact. Choices often have to take into account economic, professional, and social issues that can clash with ethical values and conceal or confuse the issue. This book is designed to help environmental professionals perceive the ethical implications of choices, identify the ethical values and principles involved with the choice, and arrive at decisions that will stand up to the scrutiny of others and fulfill personal aspirations to do the right thing.

1.2.1 Definition of Ethics

To clarify ethical issues and develop a practical approach to dealing with them, it is important to use an understandable vocabulary that reveals the differences and relationships between basic terms used in ethical problem solving. What does the term *ethics* mean? A fundamental definition in this book is: "Ethics is the difference between what a person has the right to do and the right thing to do!"[16] A person's right to do is often defined by law of the land. The right thing to do is the action taken in response to a situation that will result in the greatest benefit and the least harm to all the stakeholders. Paine presents the differences in the legal and ethical approaches to decision making elegantly.[17] While the legal tenet is to define a set of limits that must be met with a goal to prevent unlawful action, ethics provides a set of principles to guide choices to act responsibly. Our definition in this book is consistent with that. The first of two aspects of ethics involves the ability to distinguish right from wrong, good from bad, and propriety from impropriety. The second is a commitment to do what is right and proper. Ethics is an action concept, not simply something to think and rationalize about. It involves conduct and behavior. Ethics requires character and courage to meet the challenge when doing the right thing costs more than the stakeholder wants to pay.

1.2.1.1 Morals

Historically, there has never been a significant difference between the terms *ethics* and *morality*. Often the words *ethical* and *moral* are considered synonymous. Today *morals* is a term more commonly used to describe customs and traditions about right and wrong, often associated with personal values on sex, substance use, gambling, and religion. Morals can change with the times and technology. While both euthanasia and murder involve the death of one human caused by another, not to murder is an absolute ethical value, but euthanasia has changed course over the recent decades. In 1990, the U.S. Supreme Court approved the use of nonactive euthanasia. The same year, Dr. Jack Kevorkian, a Michigan physician, became infamous for educating and assisting people in committing physician-assisted suicide, which resulted in a Michigan law against the practice in 1992. Since then physician-assisted suicides have been legalized in Netherlands and Belgium (both 1993), and the U.S. states of Oregon and Washington (1994 and 2008, respectively). In 2008 author Hugo Claus, suffering from Alzheimer's disease, asked for euthanasia and died, assisted by a doctor.[18]

In this book the distinction drawn between ethical and moral is that ethical actions are absolute and moral actions may vary from time to time and society to society. Moral duties are the minimum standards of ethical conduct. Failure to properly perform a moral duty implies conduct that is wrong, unethical, or improper. There are both positive and negative dimensions to moral duties, which means moral duties require us to do certain things (e.g., be honest, fair, and accountable) as well as not do other things that might be harmful to others. Moral virtue goes beyond moral duty; moral virtues are not mandatory but something to be aspired to. Certain characteristics of conduct, such as generosity or bravery, are worthy of special praise or admiration.

1.2.1.2 Ethical and Ethically Neutral Values

Another term that is relevant to this discussion is *values*. The terms *ethics* and *values* are not interchangeable. Ethics is how a moral person should behave, but values are the inner judgments that determine how a person actually behaves. The concept of values includes a wide range of beliefs and desires that motivate behaviors. Values that pertain to beliefs about what is right or wrong do concern ethics. These are ethical values motivated by a sense of moral duty. But most values do not concern ethics. Nonethical values deal with desires or personal preferences. The desires to provide a comfortable life for your family and maintain healthy bodies are values, but are neither ethical nor unethical; they are neutral. Values regarding money, fame, status, happiness, fulfillment, pleasure, personal freedom, and being liked are nonethical values, but are not unethical. Essentially, they are ethically neutral. It is normal and appropriate to pursue these nonethical objectives as long as ethical values are not sacrificed.

Values are the beliefs or desires that motivate and guide attitudes and actions. They are the things we prize or desire the most. Most people's belief systems are based on religion, culture, family environment, laws, personal experience, and profession. They provide the basis for ranking the things we want in a way that elevates some actions based on their impact on stakeholders over others, and therefore a person's values determine how he or she will behave in certain situations. When translated into principles, these beliefs guide and motivate, and can become rules of conduct. Ethical principles are rules of conduct derived from ethical values. Values often give rise to many principles in the form of specific do's and don'ts. For example, honesty is a value that leads to these principles: tell the truth, be candid, and do not cheat or be deceptive. Values could extend toward a dress code or selection of a mate. Values are qualities to be prized, and a value system is the order in which they are prized. Because people rank their likes and dislikes and this ranking determines how they will behave in a given situation, values can conflict. The desire to be wealthy or to be kind and considerate of others can conflict with the principle of honesty. The values that are consistently ranked higher than others are termed core values, which define who we are. These are the basic do's and don'ts.

1.2.2 Making Ethical Decisions

The ethical concepts used in this book are based largely on Michael Josephson's book *Making Ethical Decisions*.[19] Josephson develops the concept of "is" versus "ought" ethics. He says that many discussions about ethics become bogged down in semantic debates, especially when ethics are taken to be relative, situational, or personal. These debates often reveal a fundamental misunderstanding of doing right and a confusion between what certain people of certain cultures actually do, i.e., the "is," and the more important question of what people should do, i.e., the "ought." "Is" ethics, or descriptive ethics, is not really ethics at all. Instead of dealing with moral obligation, "is" ethics purports only to describe operational standards of behavior—how an individual or group actually behaves—without reference to judgments of right and wrong. The use of ethics in this fashion is commonly associated with ethical relativism, which favors the considerations of the nonjudgmental agent and does not produce an ethical evaluation of actions.

"Ought" ethics are prescriptive ethics involving revelation of and commitment to principles that establish norms of behavior applicable to every person. This view of ethics describes how a person should behave based on specific values and principles that define what is right and proper. These principles may not dictate a single ethical response, but this structured approach to ethics provides a means for understanding and resolving various ethical claims. The decision-making mode we propose is based on the analytical view that ethics is not about the way things are, but about the way things should be.

Josephson argues that American culture is built on a solid foundation of ten ethical values: honesty, integrity, respect, caring, fairness, promise keeping, the pursuit of excellence, civic duty, accountability, and loyalty.[20] We believe that these ten core ethical values form the philosophical basis for ethical judgments and define the moral duties and virtues implicit in ethical behavior. Over time these virtues have evolved into the *six pillars of character*. There is an ethical dimension to every decision that can be evaluated in terms of its adherence to these six pillars. Following Josephson's approach to ethical decision making and principled reasoning, we identify six major ethical values that he contends form the core consensus on ethical values that transcend cultural and socioeconomic differences. These core values, the six pillars of character discussed in the following section, are at the heart of the ethical decision-making process.

In addition to these universal core values, people often have convictions about right and wrong, which arise from their religious beliefs, cultural roots, and political philosophy. For example, some people think drinking and gambling are wrong, others believe that it is wrong to borrow money, and still others attribute moral responsibility to dietary laws and other cultural and religious traditions and rituals. These types of moral values are treated differently in terms of ethical decision making because they vary substantially over time and between cultures, religions, or even persons of the same culture or religion. Historically, they are a source of ongoing disagreement. Although it is right for individuals with strong religious, cultural, and political beliefs to treat those beliefs with special reverence, they should be careful about imposing those types of moral values on others. This is an area where the importance of the core ethical value of respect for others, advocating tolerance, and respect for the dignity and autonomy of each person can be seen.

Furthermore, it is important to demand that all persons be treated with respect and to assert with confidence that the core ethical values of honesty and integrity are superior to dishonesty, hypocrisy, and corruption; that justice is morally superior to injustice; and that kindness and compassion are ethically superior to cruelty and indifference. A fundamental part of the core principle of treating others with respect is allowing them the widest possible latitude in matters of personal conscience. Therefore, in this context, it is not appropriate to claim that any specific religion, sexual attitude, or political philosophy is inherently superior.

The ultimate test of ethics is whether one is willing to do the right thing even when it is not to our personal benefit. When the motivation for ethical behavior is self-interest, decision making is a simple case of risk and reward. If the risk of ethical behavior is high, or the risk is low and the reward is high, then the commitment to core ethical values may get overshadowed by the temptation of beneficial outcomes. Consultants may deviate from the truth to please a client because it is good business.

An ethical approach does not always dictate a single moral course of action, but is a means of evaluating and deciding among competing options. What may seem ethical to one person may not be deemed appropriate to another person, especially one that is not directly involved with the impact of the decision on all stakeholders. "Besides believing in the importance of ethics, ethical decision-making requires sensitivity to the implications of choices and the ability to evaluate complex, ambiguous and incomplete facts and the skills to effectively implement the chosen option. While the pursuit of happiness is a basic right of free people, it is not a moral pursuit in and of itself. The morally mature individual usually finds happiness in grander pursuits than money, status, sex and mind-altering drugs. A deeper satisfaction lies in honoring universal ethical values, that is, values that people everywhere believe should influence behavior."[21] However, being ethical can also create self-esteem, the admiration of colleagues, and the respect of loved ones.

1.2.2.1 Why Act Ethically When It's Not in Our Self-Interest?

As discussed later, people are not naturally good and must learn this behavior to overcome self-interest and do the right thing to lead an ethical life. An ethical life requires that one accept a need to be ethical. Since humans are born with an instinct for self-preservation or simply selfishness, doing the right thing is not a natural human instinct. It must be developed. Let's look at the factors that account for why people perform ethically.

1. Obligations to family: Love for others and sometimes even those beyond the immediate family and care for their best interest is one factor that can cause an individual to sacrifice personal wealth and pleasure. This factor of love and care for others can become instinctive. It can develop a selfless willingness to see that other entities are not cheated.
2. Need for love from others: As people grow through childhood and develop personal relationships, the very factors that stimulate closeness to family, friends, and associates become incentives to behave ethically. For example, when your child learns from you to be kind to others and not lie, cheat, or steal, that starts off with their desire to get your love, and doing these things sensitizes them to the importance of doing the right thing even when there is no direct benefit to them.
3. Dignity: Acting ethically can develop a sense of pride that makes one be objective in reacting to situations to ensure that others are not being taken advantage of by unethical people. Thus, when facing an opportunity to take advantage of others yourself, the need to question that action becomes instinctive.
4. Respect and awe from peers: A sense of pride is one instinct that can inspire a person to do something for another person even in the event that the first person receives no benefit. For example, standing up for an associate who has been subordinated because of race, color, religion, etc., can give that person the respect of the community.
5. Recognition and respect from supervisors and regulatory authorities: When a consultant stands up to a client who is violating regulations and corrects the situation (reporting it or remediating it) even though the client resists the action, this can give that consultant confidence, especially when the situation is resolved in an equitable manner. The more successful outcomes one experiences in situations like this, the more confident a consultant becomes. The consultant gets encouraged and gains confidence to perform ethically in the next situation. People begin to respond or interact with a level of trust when the professionals perform ethically.
6. Religious training: Although it is not necessary to be religious to be ethical, and not all professed religious people are ethical, there is a value in the teachings of most religions that, if followed, are incentives to ethical behavior. "Do not do to others what you would not want them to do to you," if observed, can be a strong incentive for ethical behavior. The same could result from the "conscience" gained through philosophical thoughts and quests by atheists or agnostics.
7. Impression on others: We all crave to create good impression on others, even though we do not always demonstrate that need. When we perform ethically it creates an impression on others that can create a

warmer feeling toward us. Day-to-day behavior seldom confronts a serious ethical dilemma. But an action such as recycling or leaving a note on a parked car when we accidentally dent it and no policeman is looking can impress others, raising our self-image and inspiring us to do the right thing all the time.

8. Sensitivity to media coverage of unethical behavior by others: In government, business and industry we learn of actions of others that are roundly criticized. We read of people who have committed serious crimes who have been convicted of fraud. Seldom are these actions ones that we would like to emulate. On the contrary, we become sensitive to not act in such a manner, even when we feel that it would not be detected.
9. The boa constrictor effect: Later in the book (see Chapter 4), we describe the boa constrictor as an example of how little missteps can grow into a fatal result. We see how little crimes unpunished can lead to more serious crimes against society. Seeing this in others and reacting appropriately can also have a beneficial effect on us. We do not want to be a small-time thief growing into an armed robber like that other person.
10. Fear: The sense of fear of being caught is a significant disincentive to cheating or performing unethical acts. It is not the incentive that we would like to believe would lead to an ethical life. But realistically, we internalize this prospect, and this deterrent can have a beneficial effect on our behavior.

1.2.2.2 How to Recognize if a Situation Involves an Ethical Dilemma

But how does a person learn to recognize a situation that involves making an ethical decision? Recognizing an ethical dilemma is seldom easy. First, it is necessary to recognize situations that need a determination of right from wrong, and then it is necessary to have the courage to do the right thing even if it might sometimes hurt you or another person. This action is not always instinctual, and there is not always a correct course of action when faced with an ethical dilemma. It may be necessary to decide between various courses of action by evaluating the differences. When facing an ethical dilemma, it is necessary to recognize the dilemma, consider the elements and alternatives, and if possible, seek an opportunity to discuss the options with a trusted friend or colleague before proceeding.

We recommend environmental professionals meet regularly to discuss how to be alert to potential ethical issues in their field and how to solve the dilemmas that arise. Upon recognizing an ethical dilemma in the field, we strongly recommend it be reviewed and discussed with one or several supervisors or colleagues before any action is taken. When more than one stakeholder is involved, the right thing for one stakeholder could make another stakeholder uncomfortable. The solution to an ethical dilemma could be a compromise to minimize the impact without seriously affecting the rights or health of another stakeholder. The techniques for resolving such dilemmas are never intuitive and require training, which can be given through periodic workshops presenting hypothetical dilemmas faced by environmental professionals. Many such dilemmas are presented later in this book.

1.2.3 Six Pillars of Character

In his book *Making Ethical Decisions*, Michael Josephson defines values as the six pillars of character—trustworthiness, responsibility, respect, justice and fairness, caring, and civic virtue and citizenship—and the moral duties and virtues that flow from these primary ethical values. The six pillars of character, along with their moral duties and virtues, are

1. Trustworthiness (includes honesty, integrity, loyalty, and promise keeping)
2. Respect (includes courteousness, punctuality, and the right of self-determination)
3. Responsibility (includes pursuing excellence, competence, integrity, and self-restraint)
4. Justice and fairness (includes open-mindedness and willingness to admit error)
5. Caring (includes kindness, generosity, compassion, and avoiding harm to others)
6. Civic virtue and citizenship (includes social action, public service, and opposition to injustice)[22]

These are the core values that help people derive "the right thing to do" when resolving ethical conflicts. We discuss them below in the context of environmental ethics; one that is inherently relevant for the environmental professional.

1.2.3.1 Trustworthiness

Trustworthiness is the quality of dependability, and includes honesty, integrity, loyalty, and promise keeping. It is the most difficult test of character. When a trustworthy environmental professional makes a statement—a report, data sheet, letter, oral presentation, or conversation—the receiver should be able to rely on its accuracy and completeness. Any qualifications or misgivings should be clearly stated in order to present a complete assessment.

Honesty involves telling the whole truth; integrity takes moral courage. It is one thing to know what is right, but quite another to do the right thing. Honesty is clearly the most important element of trustworthiness. We often think of honesty as telling the truth and not stealing, cheating, misrepresenting, or altering facts. It also means telling the whole truth, clarifying the difference between fact and opinion, and revealing any misgivings one may have with the findings or conclusions. For example, when an environmental firm competes for a project, it may take credit for a previous project when the project manager and the key personnel are no longer in its employ. This claim would be dishonest unless it also states that the key people are not available, but the residual staff and systems in the firm allow the firm to maintain the capability. In consulting, "bait and switch" frequently occurs, in which an expert is presented in the proposal and less qualified alternatives are used in the implementation of the project.

In addition to telling the truth rather than using the facts to mislead, the environmental professional must be candid. This means he or she must volunteer all information the company needs. For example, a client orders a pollution control system based on the professional's design study. Afterward, that environmental professional discovers an error in the calculations that would have significantly reduced the cost. The honest approach would be for the environmental professional to reveal this situation to his or her supervisor even though it may never be discovered otherwise. The consequence is no different from the professional justifying an incorrect more expensive control by lying outright. Motivation could have come from the desire to help a friend sell that specific costlier pollution control equipment. Just as in the case of an outright lie, a concealed truth has the potential to harm.

Integrity means that knowing and doing the right thing are different. Integrity means having the personal values and moral courage to do the right thing. While honesty may be the most important attribute for trustworthiness, integrity is the most difficult. As an example, one environmental firm maintains a policy to ensure that its clients report any reportable environmental condition (e.g., a major spill), and acts with integrity by stating this policy in the terms and conditions of any contract that it accepts. The firm makes it clear that it will assume the responsibility for reporting if the client does not. Some potential clients may not be willing to accept these terms, but if the firm is committed to ensuring that violations are reported, it must have the integrity to state this principle at the outset. Many environmental firms that might hold this principle would hesitate to include such a statement in their standard contracting terms.

Loyalty requires the environmental professional to do anything within the bounds of his or her moral principles to protect public health and promote and protect the interests of the firm and client. This means doing one's utmost to protect the confidentiality of company technology, plans, and strategies, but not to the detriment of public health.

For example, a phase I property assessment must represent an exhaustive background paperwork examination and an on-site inspection. Any information that could indicate serious contamination potential must be stated and explained and a clear recommendation for further action given.

The statement should include information learned, possible interpretations, unknowns, and recommendations for further action. The recommendations for further action are critical and should reflect the perspective: "If my money were on the line for this transaction and I had to cover the cost of the recommended action, this is what I would do." Too often recommendations are made to please the client (driven by the desire for additional contracts) or to err on the side of conservatism (to compensate for one's insecurity) when advising further investigative work at the client's expense.

Josephson prioritizes loyalties. Loyalty to one's family may take priority over loyalty to one's employer within the ethical framework. For example, refusing to be away from your family for a protracted length of time on an assignment reflects the priority of loyalty to family but is not a breach of loyalty to your employer. However, accepting a higher-paying job from a competitor who wants to benefit from your knowledge of your company's proprietary technology is a breach of ethics, even if you can rationalize that the increased pay is in your family's best interest.

As an environmental professional, your moral courage is really tested when you must take action to protect public health that will surely cost you your job and may make you unemployable. In this case, your family's interests might be severely impacted by doing the right thing ethically. The coauthor of this book, while an immigrant with high aspirations, faced precisely this dilemma. On one of his very first assignments at his first job in the United States working for AECOM (formerly Environmental Research & Technology (ERT)),[23] a client wanted to respond to an agency notice of violation with an independent consultant's report that was not entirely truthful. An angry client could have cost him his job and also made him unemployable, and would have certainly jeopardized his family's continued future in the adopted country. However, he chose the ethical path and reported the situation to his superior.

Promise keeping may seem like an obvious virtue, but it encompasses contractual commitments and fulfilling the intent as well as the verbiage of a commitment. Use of extreme interpretation or loopholes to manipulate the intent of an agreement is completely inappropriate in promise keeping. The ethical environmental professional must carefully consider contractual commitments and reveal any doubts or concerns regarding the execution and scope of the agreed-upon work at the outset. Often an environmental professional firm portrays utmost confidence to a client when in fact there is a serious concern. Other environmental firms intentionally bid a low price with the aim to recover their cost and make up profit margins by overcharging on contract changes.

1.2.3.2 Respect

Respect includes courteousness, punctuality, and the right of self-determination. It means treating others, such as superiors, subordinates, clients, regulators, and contractors, in a way that respects their dignity. It includes allowing those you deal with to perform to the best of their ability. When a supervisor wants a task performed in a specified manner, the subordinate must respect his or her authority to direct it as long as it does not violate the subordinate's ethical values. On the other hand, the supervisor should allow the subordinate to exercise professional judgment and discretion in carrying out the task.

An environmental professional is often engaged to perform a task with another uninformed professional not experienced in dealing with environmental issues. That person may be a client or department manager in the same firm and may express strong opinions regarding the requested assistance. It is important that both professionals respect each other's opinions as they develop a plan of action and executing it. The environmental professional, for example, may have expertise in sampling emissions, but the client representative may better understand the process parameters that need to be controlled and recorded to obtain the maximum value from the test. In another example, a supervisor may receive assistance from a subordinate in preparing a report or a technical paper. Although the superior is presented as the principal author, it is important that the subordinate or peer is recognized as well.

1.2.3.3 Responsibility

Responsibility is comprised of integrity, accountability, the pursuit of excellence, and self-restraint. An ethical environmental professional is responsible for his own and others' actions, to the point of a certain amount of self-sacrifice. Integrity and accountability not only require the environmental professional to personally do the right thing, but also to take corrective action when others engage in unethical practices. Shifting the blame for a mistake or taking credit for someone else's work is a frequent breach of accountability. It takes moral courage to do the right thing.

A supervisor sometimes passes off an idea presented by a subordinate environmental professional as his or her own. Similarly, a supervisor may dismiss an error in a report as the carelessness of a subordinate when the supervisor failed to invoke proper quality control measures. An environmental professional's growth, as well as his or her ethical conduct, depends on his or her willingness to be accountable. In Chapter 5 we provide a case study (5.12) in which one person must take action because of the unethical practice of another.

In the environmental field, compliance costs, like taxes, are disdained, and professionals are frequently under pressure to perform at the lowest price. Budgets are tight and environmental managers try to cut corners, so cut-rate firms with minimal qualifications are sometimes employed. When a contracting environmental professional (EP) firm unethically agrees to perform a task for which it is unqualified, acceptance of the bid by the client is irresponsible, especially with the knowledge that the firm is less than competent but will likely satisfy agency requirements. The client, at a reporting facility, is accountable to the regulatory agency and to society for full compliance.

Pursuit of excellence is a moral imperative in the environmental field where technology is growing rapidly and regulations are constantly changing. Ethical environmental professionals must realize they cannot be proficient in all aspects of the field, and it is not always possible to keep pace with the state of the art in all areas. Should a task come up in an area outside one's expertise, the environmental professional should acknowledge this fact and bring in the services of a specialist in that area. The bold attitude of an environmental professional who "never met a problem he or she could not solve" could lead to trouble. Self-restraint means that sometimes it is best to admit one does not have a solution or cannot do a job. As Josephson states, "Ethical people do not adopt win-at-any-cost attitudes because if you are not willing to lose, you may have to be willing to do whatever it takes to win, even if it is unethical."[24] Even a seemingly moral cause cannot justify an unethical practice. Driving steel spikes into trees inhabited by the spotted owl to protect an endangered species was an outrageously immoral act that resulted in the death of loggers attempting to harvest those trees. One unethical practice never justifies another. For instance, as a doctor I may be morally opposed to abortion or euthanasia, but denying these procedures for my patients or imposing my views on my peers is grossly irresponsible, especially if the practice in question has been adopted as an acceptable practice in the medical profession.

1.2.3.4 Justice and Fairness

The principles of justice and fairness encourage one to be objective and impartial. However, deciding what is fair can be difficult. For instance, if an environmental coalition is advocating the closure of a factory employing a major segment of a town's workforce because of its proximity to schools, hospitals, and residences, it is easy to perceive the inherent biases that might motivate environmental professionals employed by the plant versus those hired by the coalition.

The principles of justice and fairness require the ethical firm to voluntarily do whatever is required to protect public health. But suppose the plant existed before the homes and schools were built within the impacted neighborhood, before anyone realized that the pollution released from the plant is toxic. The company claims that if the level of control required to protect the public health is installed, the plant cannot compete favorably in the market for its product. The plant management does not wish to be uprooted even if the company could afford to relocate. Should the town move? What action is fair? Is it possible for the two groups to conduct an objective analysis and reach an equitable solution? In the ideal world, they would. For instance, it may be more cost-effective to relocate the school with funding from the plant.

1.2.3.5 Caring

Caring requires minimizing the negative impacts of a decision on all concerned. Environmental professionals must carefully identify all stakeholders affected by a decision, consider alternative solutions, and choose the one that will accomplish the objective with the least disruptive impact.

For example, an environmental professional discovers a reportable infraction of water discharge regulations during an audit. The client directs the professional to refrain from mentioning the problem in the audit report because he will have the matter corrected. Legally, the responsibility to report the infraction is with the client's firm. Ethically, the environmental professional feels the matter must be reported. However, if that environmental professional takes this initiative, the repercussions could be extensive for both the environmental professional's and the client's firm. A more caring solution would be to convince the client or an executive in his firm to face the situation and deal with it. Only if all means to have the client's firm report the infraction have failed should the environmental professional consider notifying the regulatory agency.

1.2.3.6 Civic Virtue and Citizenship

Civic virtues and citizenship give the ethical environmental professional an obligation to the community that extends beyond self-interest. Civic virtues emphasize social contributions, such as running for public office, working for candidates or issues, accepting appointments to public office, and giving time and money to social causes. An environmental professional who is a responsible citizen uses his or her special knowledge on community planning and environmental advocacy committees.

Environmentalist groups often begin with the best intentions, but lack knowledge or misdirect their advocacy. Through participation in grassroots efforts, environmental professionals can help educate and steer these groups. The coauthor, for instance, has experienced numerous situations in which regulatory agencies, environmental activists, and industry advocates have been able to save a significant amount of society resources with creative and innovative solutions. For instance, while leading a Superfund response group on behalf of the industry at a Superfund site, the coauthor advocated and implemented natural attenuation, as an alternative to the proposed dig, move, and treat option, which lessened risks to public health and also saved a considerable amount of social resources.

1.3 Ethics, Philosophy, and Religion

In the previous section we touched on the ten core ethical values that form the philosophical basis for ethical judgments and define the moral duties and virtues implicit in ethical behavior. We also discussed the six pillars of character, which evolved from the ten values formulated by Michael Josephson.[25] This approach to ethical conduct does not assume any religious background in justifying the recommended actions—one does not have to believe in a higher authority to lead a moral life.

Also, as noted earlier, people develop values and views about right and wrong based on religious faith, cultural background, and political leanings. Therefore, knowledge of religious values can serve as a guide to appreciating the qualities behind the pillars of character. It is important for the reader to understand that many of those same qualities are also derived from the golden rule: "Do to another person what you would want him or her to do to you." Finally, learned behavior, especially in the formative years, goes a long way in defining character and its consequent impact on ethical or unethical behavior. As discussed earlier, character, which is formed at an early age, defines the extent to which people will go to achieve an objective. Gandhi always held that a prerequisite to building a strong character is to always be cognizant that the means is as important as the end goal. In this section we will discuss values derived by humans through learning from religious faiths and atheism, values that may be adopted by the environmental professional to help form a sound basis for doing the right thing.

Hundreds of publications, books, magazines, newspaper columns, and papers have been written on ethics and morality, dealing with issues ranging from personal human behavior to governmental actions. Many of these publications are written by scholars who define values that lead to doing the right thing, while others have a specific religious basis referencing specific passages from holy writings. Most humans naturally draw their sense of values from multiple sources, including religion and philosophy, and often these have a synergistic effect. They cement convictions about what are the right things and a commitment to do the right thing. Some others may desire a further quest into the basic source of these ethical values. With that in mind, a list of pertinent examples of these publications is included in the endnotes.[26] These publications discuss the values that one should develop to lead a good life and the religious and philosophical basis for those values, and the moral choices that one may confront and the values that lie behind those choices. They are about innocence, guilt, right and wrong, and what it means to live a good or bad life. They discuss the dilemmas of life, death, sex, violence, and money. They explore human virtues and vices, rights and duties. They discuss matters concerning the sanctity of human life, such as abortion and euthanasia.

1.3.1 Religious Values

Religious values are those specified by a godly figure and subject to the interpretation of the clergy, or by philosophers of particular cultural groups. There are many different faith groups, but this section covers the five major ones: Christianity (>2 billion), Judaism (~12 million), Islam (>1.5 billion), Hinduism (>1 billion), and Buddhism (~1 billion). It also covers the group of agnostics and atheists, which numbers close to a billion people. However, it must be noted that there are many nativist religions (Shinto of Japan) or the very broad array known as Confucianism, mostly found in China and its colonies. Most nativist religions, such as found in both Central and South America, hold a set of religious beliefs that are very emphatic to the ethical treatment of the environment. In fact, one could argue that nativist religions are almost by definition focused on the ethical treatment of the environment. The universal moral statement of all religions is "do unto others what you would have them do unto you," or "what is hateful unto you do not do to your neighbor." Knowing how to act appropriately is often not a simple matter and can require years of study. As U.S. president Lyndon Johnson once said, "A president's hardest task is not to do what is right but to know what is right."[27]

1.3.2 Christianity

Followed by about two billion people, Christianity has the broadest following of any religious faith. It is based on the Bible, both the Tanakh (also commonly known as the Hebrew Bible or the Old Testament) and New Testaments. Tanakh is also the basis for the Jewish faith, which is discussed next. Christianity accepts the moral basis of the Ten Commandments and the Tanakh, or the Hebrew Bible, as part of its own scriptures. Early in the history of the Christian church it was agreed that its members should not be required to keep all the rules of the Hebrew Bible. Most of Christianity (not all) is trinitarian and emphasizes faith in God, Jesus Christ, and the Holy Spirit, rather than obedience to religious rules as the basis of its way of life. It interprets the rules of the Old Testament in the light of the life and teachings of Jesus Christ. A significant feature of the New Testament is that while Jesus was prepared to set aside the detailed requirements of the law, he insisted that in doing so he was not setting the law aside, but rather fulfilling and completing it. Also, the Catholic church teaches that "heaven is the ultimate end and fulfillment of the deepest human longings, the state of supreme, definitive happiness."[28] This has allowed Christian morality a measure of interpretative flexibility.

Christianity is in general less rule based than Judaism, even where it shares the same values. One could easily argue the opposite: some denominations of Christianity are highly rule bound. There are elements of Judaism, which is not a uniform faith by any means, that are highly interpretive of the underlying law (Torah). Nevertheless, Christians and Jews hold a common core of moral tradition. Sources of authority for Christian morality vary from one Christian denomination to another. But in general they include the authority of the church (inspired by the Holy Spirit), including the pope, who is the bishop of Rome, the authority of the scriptures, human reason (used by applying a "natural law" basis for morality), conscience, and the direct inspiration of the Holy Spirit. However, the authority of the Catholic church is not accepted by most of the Protestant movement. Nevertheless, the fact that there are a number of different sources of authority is significant. Consequently, at the time of the Reformation in the sixteenth century, the authority of the church was challenged on the basis of the authority of the scriptures.[29]

Incorporating human reason and conscience as sources of authority for moral values in Christianity is significant in reexamining the relevance of particular teachings to changing cultural and social circumstances. Conscience is the inner conviction that something is right or wrong. It is mentioned in considering the various ethical theories and is found in one form or another in both religious and secular ethical debates. From a religious point of view, conscience may be seen as the voice of God within the individual, or more generally as the operation of a fundamental moral sense found within everyone. From a secular perspective, conscience may be seen as coming from internalized values that have been instilled by society or training. It can be closely related to psychology, and particularly to Freud's superego, the part of the mind that keeps the basic animal instincts in check.

There is a long tradition of moral thinking that accepts conscience as giving moral authority. Early philosophers believed that humankind had a natural sense of benevolence, and repugnance at the sight of the suffering of others, which led to a "moral sense" and conscience.[30] Thus, emotion and conscience, rather than reason, were the starting points for morality. Aquinas argued that there were two ways to act wrongly: one was to go against what one rationally believed to be right, and the other was to act against one's conscience.[31]

1.3.3 Judaism

Jewish morality is based on the first five books of the Tanakh, or the Hebrew Bible, known as the Torah, which means "teaching" or the "law." The Torah is the collection of the first five books and distinctly a subset of the Tanakh.

The interpretation of the Torah was developed over centuries by the sages as each law was applied to a new situation. The many layers of tradition that surround the original Torah represent the cumulative experience of generations. The Torah includes many rules for ethical and social matters; the most well-known of these rules are the Ten Commandments. The Torah liturgy was summed up by Rabbi Hillel in the first century BCE. His statement of the golden rule is: "What is hateful to you do not do to another."[32]

Around 500 to 600 CE a complex encyclopedia of rules and traditions known as the *Talmud* was completed. It comprises two books, the first of which is the *Mishna*, a collection of oral traditions believed to be originally given to Moses by God along with the written text of the Torah. The second book of the Talmud, known as the *Gemara*, was developed over the years and presents further commentaries on the Mishna. Observant Jews are required to follow these laws in the correct way, known as the *halakah*, or "path." Different groups within Judaism vary in the level of strictness with which they put these rules into effect. Some take them literally in every detail, and others follow the general principles, adapting as necessary to some perceived requirements of modern life.

In terms of morality, the Jewish approach is one of rules. Reason and conscience have their part to play, but this is mainly in terms of interpretation and application. It is important to emphasize that the rules are not arbitrary. They are seen as having the authority of God, given to mankind for its benefit. There is a real sense that in keeping the rules of a person's religion, a follower of the Jewish faith is not simply adopting a particular style of life, but entering into a community and a tradition. Many who may not practice Judaism religiously, as well as many nonbelievers, belong to Jewish social groups in which the ethical aspects of the faith are discussed and practiced, such as kindness and charity.

1.3.4 Hinduism

Followed by over a billion people, Hinduism is the oldest living major religion that grew organically and covers a broad spectrum of monotheists, polytheists, and atheists. "Hinduism is not a religion in the usual sense. It is a civilization based on a simple metaphysical insight about the unity of the individual and the universe and has self-development as its objective."[33] Hinduism is best described as a way of life. You are a Hindu not simply because you worship one way, but because you accept a set of principles or code of practice for life in the world. In Hinduism, your purpose of leading a moral life is not to make god happy or for you to avoid hell and go to heaven instead. In Hinduism, achieving perfection is the supreme purpose in life. This includes aspiring to and achieving ambitions in life, and doing so with a sense of duty that is not driven by a selfish motivation of personal gain or by fear of punishment.

Hindu scriptures are comprised of the four Vedas (composed about 1500 BCE), which had a vast influence on Buddhism, Jainism, and Sikhism, and the three canonical texts: *Upanishads* (written between 800 and 400 BCE)—a philosophical inquiry, which also addresses how the soul unites with the ultimate truth as well as the doctrine of karma; *Brahmasutras*, which sets a logical order; and *Gita*, which gives a practical context. While *Upanishads* focuses on monism, *Gita* integrates dualism and theism.

In addition, *Mahabharata*[34] and *Ramayana*[35] are epics (written around 500 BCE and probably the longest poems in any language) based on oral traditions that provide moral and spiritual themes in a story format. These were written to make the scriptures understandable to the general population. The battlefield setting of *Gita*, within *Mahabharata*, where Krishna counsels Arjuna, metaphorically depicts the ethical struggles of human life.[36]

Hinduism believes, per the Rig Veda (1500 BCE), that for harmony with the cosmic order that regulates all movements, from bosons to multiverses in today's context, humans should have a similar moral order. *Dharma* is the cosmic contract of individual rights and duties. It assigns a specific place to all animate and inanimate constituents of the cosmic family. It goes beyond humans and defines the purpose of being even for inanimate objects, such as water, which exists to quench thirst. It also imposes obligations on all humans toward self, family, and society, as well as toward all fellow creatures and all nonliving objects in the cosmos.

In the context of ethics that relate to sustainability and environmental conservation, *dharma* in Hinduism provides one of the best definitions of values to draw from. Rooted in the Sanskrit word *dhr*, *dharma* means sustain. It encompasses principles, rules, morals, and duties of individuals, society, and the ecosystem.

> If *nirvana* or *moksha* is full divinity, *dharma* is what is feasible under human conditions.... In the Hindu legal literature, *dharma* conveys the same meaning as ethical, reasonable and equitable used in western legal literature. The concept of *dharma* is more profound than the prevailing broader concepts of duty and ethics. Individual duties and responsibilities take precedence over individual rights and privileges.[37]

In social life, dharma represents just and equitable laws, suppression of evil, and promotion of virtues, such as compassion, honesty, justice, and responsibility. It is the formula for "doing the right thing" and getting the job done right. It is the right thing to do under the given scenario. As Das[38] clarified, the definition of moral duty in Hindu philosophy is contextual.

The Hindu code of practice provides a description of right conduct or duty, and this in turn provides the context that determines what is morally right. Your particular duty depends on what you do for society and in which stage of life you are. Hinduism recognizes that different people have different abilities and are able to take on different responsibilities. It does not hold that society is uniform, or that an equal distribution of goods or equal rights will in itself produce happiness or justice. Hindu society is classified, based on the duties performed, into castes or *varnas*. *Brahmins* have a duty to pursue education and philosophy and enrich the society. *Kshatriyas* have to be strong and are charged with protecting the safety and security of the society. They are often part of the royalty, military, or political leaders. *Vaishyas* are expected to be business savvy with a commercial mind, and be entrepreneurial and promote trade and business, with the explicit duty of making goods and services available to the society. *Shudras*, who are less skilled in intellectual pursuits, are expected to serve the society through manual labor. It is important to note that the caste system was not meant to promote inequality, but rather to value differences. People were to be given respect, whatever their role in society. Unfortunately, for a period it degenerated to a point where the castes were defined by birth rather than skills and duties.[39]

Hinduism also defines four stages in life, known as the *ashramas*. Each of these stages has moral duties and each implies certain moral choices as follows. The first stage is that of a student. At this stage a person is expected to work hard, show respect to parents and teachers, and develop self-discipline. Students (*brahmacharya*) are expected to abstain from sex, alcohol, tobacco, and drugs. The next stage is that of a householder (*grihastha*); during these middle years of life a person is expected to be involved with marriage, family, and career. The social rules for this stage are the most lenient. At the retired (*vanaprastha*) stage, traditionally from the time of the arrival of a grandchild, it is expected that both husband and wife may start to stand back from family and business concerns. At this point they can choose to hand over control of a family business to the next generation. The final stage is that of an ascetic (*sanyasa*), where in the last years of their life they can choose to divest themselves of all possessions and devote themselves to the quest of meaning of life and spirituality. Most remain in the retired stage, however, and continue to live with their families.

In summary, Hindu moral conduct revolves around *dana*, *daya*, and *dama*. *Dana* enthuses one to give others their due share, *daya* urges one to treat all of God's creatures with kindness and compassion, and *dama* counsels control of passions, so that desires may not exceed their limits. In the context of the planet and the environment, this has been rather aptly conveyed by Gandhi, when he said, "The earth provides enough to satisfy every man's needs, but not every man's greed."[40]

1.3.5 Buddhism

Followed by over half to nearly one billion people, Buddhism was started around 500 BCE by Gautama Buddha, a spiritual teacher in the Indian subcontinent. Unlike most other religious traditions, Buddhism is not based on a belief in God, but offers a practical set of guidelines for spiritual development. A Buddhist's objective in life is to overcome the suffering caused by greed, hatred, and illusion, and to develop in a way that leads to peace, joy, and insight. Buddhists believe that all ethically significant actions have consequences. What you do today will affect the sort of person you become tomorrow. This is not in the sense that there are any externally imposed laws to that effect, but rather just that actions have their natural consequences. Those who are foolish remain trapped in a cycle of craving and dissatisfaction. Buddhists do not refer to actions as good or bad, or right or wrong, but as skillful or unskillful. This implies that an action is assessed in terms of the intention of the person who does it rather than in terms of any external absolutes. An action is skillful if it is the result of insight and loving kindness; it is unskillful if it is the result of ignorance, greed, or hatred. Unskillfulness is shown both in the action and in the results that arise from it.

Buddhist moral guidelines are summed up in five precepts. The first precept is not to destroy a life—avoid killing other humans, and also (for many Buddhists) follow a vegetarian diet, so as to minimize the suffering of other species. It also implies avoiding a negative attitude toward life, such as denigrating people or situations, instead cultivating an attitude of loving kindness. The second precept calls upon one not to take what is not given—avoid dishonesty, which includes obtaining material goods in a way that may be honest but implies craving—and instead cultivate an attitude of generosity. The third precept advises one not to indulge in harmful sexual activity—avoid use of sexuality in such a way as to harm people. Although many Buddhist attitudes are shaped by the cultures within which they developed, in general, Buddhism takes a liberal and positive attitude toward sex. However, those who seriously aspire to follow the spiritual path may find that they are able to gain personal fulfillment without sexual activity. This precept is sometimes taken in a broader sense of not harmfully overindulging any of the senses. The fourth precept directs that one not speak falsely—avoid lying—which includes deliberately giving a false impression. The final precept suggests that one not take those things that cloud the mind—stay alert to increase one's own awareness and be sensitive to your own and others' feelings and responses. This cannot happen if the mind is dulled by alcohol or drugs, and so it recommends that they be avoided. Unlike some religions, there is no absolute prohibition.

Buddhists are also required to cultivate four mental states, known as the *brahma viharas*, which may be translated as the "the resting place of the divine spirit."[41] These are

Love—toward all creatures and oneself

Pity—compassion for all who suffer

Joy—an unselfish sharing in the happiness of others

Serenity—freeing oneself from anxieties of success or failure, and being equal minded in dealing with other people

1.3.6 Islam[42]

Islam started prior to 600 CE and is the youngest of the major religions in practice today, yet it has over 1.5 billion followers. Islam is unique because of its timing in history; it could build on the teachings of Christianity and Judaism. It is seen by its followers as a complete and universal version of monotheistic faith revealed previously in various places and times by many, including Abraham, Moses, and Jesus.

Islam is a way of life and teaches its followers (called Muslims) how to act and be in every aspect of life. There are five basic pillars or components of the religion that dictate the life of a practicing follower.[43] The first pillar is that Muslims believe in only one God, Allah (Arabic for God), who has no partners or family and who created the universe. The second pillar is prayer, which provides the believer with several opportunities throughout the day to thank God for all of his blessings and ask God for help in all other endeavors. The third pillar, fasting, is an opportunity for self-reflection and improvement. While fasting and feeling the pains of those that do not have enough, Muslims are forced to reflect on their blessings and give back. Charity, which is the fourth pillar, is the foundation of all of the teachings and does not refer merely to monetary donations, but includes respect and love for all mankind. Respect and love for all mankind is a big facet of Islam, since it holds that all human beings are descended from Adam and Eve and are thus brothers and sisters, and as such must care for each other. Harming another human being is strictly prohibited. The golden rule—"Do unto others as you would like done to you"—is present in all traditions of Islam. The fifth and final pillar is the Hajj, the sacred pilgrimage to Islam's most sacred and holiest site, the Kaaba (the Holy Mosque in Mecca, Saudi Arabia), which provides Muslims an opportunity to cleanse and purify their soul. Muslims are only required to perform the Hajj once in their lifetime.

The book of Islam, the Qur'an, identifies Moses and Jesus among the many prophets and messengers that God sent to mankind throughout the ages to educate people and spread the word of God. Though they did not possess any superhuman powers, God enabled them with miracles in order to capture the attention of their people and provide them with the ability to intrigue and convince the masses. God ended the lineage of messengers sent to mankind with the last and final prophet, Muhammad, who was responsible for spreading the teachings of Islam and leading a life of example. His words are known as hadith, and his actions are called the sunnah.[44] Muslims aim to live their lives as the Prophet Muhammad had done.

The key characteristic of Muslims that practice Islam is accountability. At the end of it all, Muslims believe in the day of judgment, when all humans will be judged by God according to their actions in life and sent to either heaven or hell. Hence, this world is not the ultimate goal, but rather success in the hereafter or life after death. Therefore, most Muslims believe in brotherhood, peace, equality, love, and friendship.

The word *Islam* means submission. Muslims believe that the purpose of life is to love and serve God, or Allah, and that most things in the world submit naturally to God, the source of life. Humans are the exception, as people can either choose to submit to God and thereby live in a natural way, or refuse to do so.

Muslims believe that Allah revealed his will for humankind through various prophets, both Christian and Jewish, but his final revelation was given by the Prophet Muhammad. The Qur'an and the hadith (depending on the major variant of Islam) are the primary sources of Muslim authority. Both are recitations from the Prophet Muhammad and form the basis of *shariah*, a set of natural laws created by God. *Shariah*[45] determines everything in the universe, from the movement of the planets to the details of how people should act. It forms not only an ethical system and legal framework, but also an entire way of life.

As they apply shariah to present-day situations, Muslims use the principles of analogy—linking a present issue to a decision found in the hadith or Qur'an. Where there is a doubt about what is right, a gathering of Muslim scholars is called to reach a decision on a point of law. Because Muslims believe that their community can never agree together on something that is wrong, the decisions of this council of scholars are authoritative.

1.3.7 Enlightened Self-Interest, Agnosticism, and Atheism[46]

Up to this point we have discussed the influence of several major religions on human ethical behavior, but it is estimated that there are over one billion people in the world who do not follow any religion. Do you have to be religious to be ethical? Actually some of the most ardent proponents of ethical behavior do not have a personal belief in God. When an atheist or agnostic approaches the problem of finding natural justifications for human morals and establishing a nonsuperstitious basis for ethical behavior, at first it might appear as though it is not much of a problem at all. Humans naturally desire love, seek beauty, and are thrilled by the act of creation. Complexity creeps in when we examine traditional moral codes that do not arise out of necessity. These moral mandates are seen as the result of vain attempts to accommodate human needs and nature to the dictates of demons and deities. Ethicists like Peter Singer have argued that ethical behavior, regardless of the agent, is a result of the same causes, is regulated by the same forces, and has nothing to do with the presence or absence of religious belief. Often, Aristotle is cited to describe unethical behavior as when man's rationality is overcome by his desire. The same set of universal rules governs the behavior of all human beings. Even among the people of religious faith, fear of hell or enticement of heaven is the not the key motivation for ethical behavior. Stated in an earlier section on Hinduism, this applies across most major religions.

Another Greek philosopher, Plato, believed that people cannot depend upon the moral demands of deities. Plato asked whether "the commandments of a god were good simply because that god had commanded them or because the god recognized what was good and then ordered the action accordingly. If you believe that something is good simply because a god has commanded it, anything could be considered good. There would be no way of predicting what in particular the god might desire next, and it would be entirely meaningless to assert that God is good."[47] Stoning adulterers would be just as likely to be good as would the principle of philanthropy.

If god is not the source for the gold standard of what is "good," what criteria could we apply to choose what is right from the alternatives?

Enlightened self-interest provides a good starting point to develop a basis for what is right and ethical. An analogy could be drawn from economics and the markets. If stealing were not illegal, there would be no sustainable ownership, be it a car today or, for that matter, even a wife in the cavemen days. Likewise, if one were to practice taking every object of one's interest in a selfish way, pretty soon there would be pandemonium. This experience would quickly lead the uninitiated to enlightened self-interest. This offers the beginnings of respecting each other's rights and the value of social rules.

Next is the development of the concept of cooperation, which provides synergy, where one realizes that one plus one is larger than two in what could be accomplished. It can be said that one man with a rock cannot kill a large wild animal for dinner. But a group of men or women, with lots of rocks, can drive that animal off a cliff. Even after dividing the meat up among them, they will still have more to eat than they would have had without cooperation. On the other hand, cooperation can work both ways. If one person cooperates with several others to kill wild animals, and then they drive that person away from the kill and eat it themselves, the person will quickly take his or her services elsewhere and leave the selfish participants to hunt without assistance. Also, man is a social animal. Humans are generally happier in the company of their fellow creatures than alone. Because they are emotionally malleable, as they practice enlightened self-interest they usually will be wise and choose behaviors that will make others happy and willing to cooperate with and accept them.

Justice involves reciprocity and the problem of determining what is fair. If a worker cooperates with a farmer by tilling his field of corn, how much does the farmer owe the helper at harvest time? Justice involves cooperation at maximal efficiency, with increasingly desirable benefits. Therefore, the idea of enlightened self-interest involves a desire for justice. Fair trade and changing social contracts between businesses and societies, and even between countries, are results of this evolving principle. Justice and cooperation can produce a winning team. But without justice and cooperation, they do not even make it to the competition.

Haidt succinctly captures the basis of human society and collaboration in his book *Righteous Mind*.[48] There are four basic pillars of collaboration in human society: care, cooperation, compensation, and choice. In today's context, collaboration in human society relates to the formation of political groups. Haidt describes how each of the four moral foundations of politics was borne out of necessity: (1) The value of care rather than harm comes from caring for the vulnerable human child. (2) Adapting to cooperation, while preserving self-interest, forms the basis of the value of fairness and not cheating. (3) The need to expand and sustain coalitions calls for rewarding loyalty and punishing betrayal. At this stage groups recognize differences and the social value of a hierarchical structure based on ability and skills.[49]

The next logical step in the development of what is right is the recognition that different people have different abilities and are able to take on different responsibilities; this is consistent with the caste system in Hinduism. This recognition has led to the formation of leaders, peers, and followers, and status quos relevant to expected behavioral norms. Finally, humans, like rats and roaches, are omnivores who must both explore and be wary, a fact from which the environmental movement draws heavily. Because they are omnivores, they can adapt to different environments and choose what to eat. They, by nature, are more exploratory and also more wary of changes in their surroundings. Some activists find industrial capitalism to be a degradation of nature, a deliberate choice made by a group, and hence repulsive.

Moving further, because the human nervous system is impressionable, people are able not only to fall in love at first sight, but also to love objects and ideals as well as people, and love them with variable intensities. Like a hummingbird drawn to a flower, humans are pulled forward by a desire for love. Unlike the hummingbird's desire, however, a human's love can be shaped by experience to a considerable extent and can be modified by education. A major aim of enlightened self-interest is to give and receive love. As a general, though not absolute, rule, people must choose behaviors that will likely bring them love and acceptance and avoid behaviors that will not.

Finally, another aim of enlightened self-interest is to seek beauty in all its forms, to preserve and prolong inner beauty and the splendor of the world outside. Beauty and love are just different facets of the same jewel. Love is beautiful, and we love that beauty. The experience of love and beauty, however, is a passive function of the mind. The joy that comes from creating beauty is great. Whenever a person transforms the raw materials of existence so they are better than they were when they were found, that person has been creative and can enjoy the fruits of success.

When approached from the perspective of enlightened self-interest, the task of moral education is not to memorize great lists of do's and don'ts, but to help people foresee the consequences of the actions being considered. What are the immediate and long-term rewards and drawbacks of the acts? Will an act increase or decrease a person's chances of achieving love, beauty, and creativity? A company's code of ethics may list do's and don'ts that may be read periodically and acknowledged in writing, but this process is not nearly as effective as an ethics workshop where the implications and consequences of these moral principles are discussed and related to ethical dilemmas the participants are likely to encounter.

Endnotes

1. Beth Young, Celine Suarez, and Kimberly Gladman, *Climate Risk Disclosure in SEC Filings*, The Ceres Corporate Library, last modified June 2009, http://www.ceres.org/resources/reports/climate-risk-disclosure-2009.
2. M.A. Delmas and V.C. Burbano, "The Drivers of Greenwashing," *California Management Review* 54, no. 1 (2011): 64–87.
3. Testimony of Chairman Alan Greenspan, while presenting the Federal Reserve's Monetary Policy Report, Federal Reserve Board, July 16, 2002, archived from the original on June 7, 2011, retrieved July 13, 2011, accessed December 2012, http://en.wikipedia.org/wiki/Alan_Greenspan.
4. Rachel Carson, *Silent Spring* (Boston: Mariner Books, 2002), accessed December 2012, http://www.goodreads.com/work/quotes/880193-silent-spring.
5. D. Callahan, *The Cheating Culture: Why More Americans Are Doing Wrong to Get Ahead* (New York: Harcourt Publishers, 2004).
6. William J. Bowers, *Student Dishonesty and Its Control in College* (New York: Columbia University Bureau of Applied Social Research, 1964).
7. Callahan, *The Cheating Culture*.
8. Lynn Sharp Paine, *Value Shift—Why Companies Must Merge Social and Financial Imperatives to Achieve Superior Performance* (New York: McGraw Hill Professional, 2003), in Preface.
9. Callahan, *The Cheating Culture*.
10. Richard Lacayo and Amanda Ripley, "They Took Huge Professional and Personal Risks to Blow the Whistle," *TIME*, December 22, 2002, http://www.wanttoknow.info/021222time.personofyear.
11. Wikipedia, s.v. "Henrik Ibsen," last modified September 17, 2012, http://en.wikiquote.org/wiki/Henrik_Ibsen.
12. Martin L. King, "Rev. Dr. Martin Luther King, Jr. Quotes," accessed December 2012, http://mlkday.gov/plan/library/communications/quotes.php.
13. Lynn Stout, The *Shareholder Value Myth—How Putting Shareholders First Harms Investors, Corporations and the Public* (San Francisco: Berrett-Koehler Publications, 2012), p. 1.
14. Danielle Ivory, "BP Temporarily Banned from Contracts with U.S. Government," *Bloomberg News*, November 28, 2012, http://www.bloomberg.com/news/print/2012-11-28/bp-temporarily-suspended-from-new-contracts-with-u-s-government.html.
15. Ronald R. Sims, *Ethics and Corporate Social Responsibility* (Westport, CT: Praeger Publishers, 2003), p. 147.
16. Justice Potter Stewart, accessed May 1, 2013, http://thinkexist.com/quotes/potter-Stewart/2.html.
17. Lynn Sharp Paine, *Venturing beyond Compliance: The Evolving Role of Ethics in Business* (New York: The Conference Board, 1996), pp. 13–16.
18. Wikipedia, s.v. "Voluntary Euthanasia," accessed December 2012, http://en.wikipedia.org/wiki/Voluntary_euthanasia#Modern_history.
19. Michael S. Josephson, *Making Ethical Decisions* (The Josephson Institute of Ethics, 2002), accessed December 2012, http://josephsoninstitute.org/MED/MED-2sixpillars.html.

20. Michael Josephson, "Making Ethical Decisions," accessed December 2012, http://www.sfjohnson.com/acad/ethics/Making_Ethical_Decisions.pdf, p. 1.
21. Ibid., p. 6.
22. Ibid., p. 1.
23. Formerly ERT, now AECOM, one of the largest environmental consulting companies in the world. Accessed December 2012 at http://www.aecom.com/.
24. Josephson, "Making Ethical Decisions," p. 21.
25. Ibid.
26. Georgia Harkness, *Christian Ethics* (Nashville, TN: Abingdon Press, 1957); Joseph Telushkin, *A Code of Jewish Ethics* (New York: Crown Publishing, 2006); Mel Thompson, *Ethics* (New York: McGraw Hill, 2000); Stephen Asma, *Buddha: A Beginner's Guide* (Charlottesville, VA: Hampton Roads Publishing, 2008); Jean-Francois Revel, Matthieu Ricard, John Canti, and Jack Miles, *The Monk and the Philosopher: A Father and Son Discuss the Meaning of Life* (New York: Random House Publishers, 1999); Bansi Pandit, *Explore Hinduism* (Leicester, UK: Heart of Albion, 2005); Gurcharan Das, *The Difficulty of Being Good on the Subtle Art of Dharma* (London: Oxford University Press, 2009); Peter Singer, *How Are We to Live? Ethics in an Age of Self-Interest* (Hong Kong: Mandarin Press, 1995); Charles Colson and Nancy Pearcey, *How Now Shall We Live?* (Carol Stream, IL: Tyndale House, 1999.)
27. "President Lyndon B. Johnson's Annual Message to the Congress on the State of the Union January 4, 1965," University of Texas, last modified June 6, 2007, http://www.lbjlib.utexas.edu/johnson/archives.hom/speeches.hom/650104.asp.
28. Wikipedia, s.v. "Heaven (Christianity)," last modified November 1, 2012, http://en.wikipedia.org/wiki/Heaven_%28Christianity%29#cite_note-11.
29. Theopedia, an Encyclopedia of Christianity, s.v. "Protestant Reformation," accessed December 2012, http://www.theopedia.com/Protestant_Reformation.
30. Stanford Encyclopedia of Philosophy, s.v. "Medieval Theories of Conscience," last modified July 7, 2011, http://plato.stanford.edu/entries/conscience-medieval/#BM3.
31. Theopedia, "Protestant Reformation"
32. Wikipedia, s.v. "Hillel the Elder," last modified November 20, 2010, http://en.wikipedia.org/wiki/Hillel_the_Elder.
33. Das, *Being Good*.
34. Wikipedia, s.v. "Mahabharata," last modified December 16, 2012, http://en.wikipedia.org/wiki/Mahabharata.
35. Wikipedia, s.v. "Ramayana," last modified December 16, 2012, http://en.wikipedia.org/wiki/Ramayana.
36. Wikipedia, s.v. "Bhagavad-Gita," last modified December 17, 2012, http://en.wikipedia.org/wiki/Bhagavad_Gita.
37. Pandit, *Explore Hinduism*.
38. Gurucharan Das (author of "India Unbound" and former CEO of Procter and Gamble, India), private conversation with coauthor's wife, Janaki Ramanan, at the release of his book *Difficulty of Being Good* on October 14, 2010, in Chicago.
39. Wikipedia, s.v. "Caste System in India," last modified December 3, 2012, http://en.wikipedia.org/wiki/Caste_system_in_India#Modern_status_of_the_caste_system.

40. Govind Singh, "Mahatma Gandhi—A Sustainable Development Pioneer," Eco Localizer, last modified October 14, 2008, http://ecoworldly.com/2008/10/14/mahatma-gandhi-who-first-envisioned-the-concept-of-sustainable-development/ in http://www.mkgandhi.org/articles/environment1.htm.
41. Dhamma Encyclopedia, s.v. "Brahma Viharas," accessed December 2012, http://www.dhammawiki.com/index.php?title=4_Brahma_Viharas.
42. John L. Esposito, *Oxford History of Islam* (New York: Oxford University Press, 2000).
43. Sue Penney, *Islam* (Oxford: Heinemann, 1999), p. 14.
44. Jonathan A.C. Brown, *Hadith: Muhammad's Legacy in the Medieval and Modern World* (Oxford: Oneworld, 2009), p. 6.
45. Janin Hunt and André Kahlmeyer, *Islamic Law: The Shariah from Muhammad's Time to Present* (Jefferson, NC: McFarland and Co., 2007), p. 1.
46. This section includes concepts rephrased and rearranged from the following references: Peter Singer, *Writings on an Ethical Life* (New York: Ecco Press of Harper Collins, 2000); Paul Kurtz, *Forbidden Fruit: The Ethics of Humanism* (Amherst, NY: Prometheus Books, 1988); Edward Ericson, *The Humanistic Way: An Introduction to Ethical Humanist Religion* (New York: Continuum, 1988); Hazel Barnes, *An Existential Ethics* (Chicago: University of Chicago Press, 1985); A.C. Grayling, *Meditation for the Humanist: Ethics for a Secular Age* (London: Oxford Press, 2002); and F.R. Zindler, *The Probing Mind*, accessed December 2012, http://atheists.org/content/ethics-without-gods.
47. Zindler, *Probing Mind*.
48. Jonathan Haidt, *The Righteous Mind* (New York: Pantheon Books, 2012), p. 128.
49. Ibid., p. 131.

2

Environmental Ethics and Public Policy

2.1 Human Beings and the Environment

As a society progresses it often faces a conflict between economic benefits and environmental degradation. Many environmentalists oppose drilling for oil and cracking shale to produce natural gas because of the potential impact on local natural resources. On the other hand, banning DDT, a chemical known to be hazardous to human health, is now believed to have allowed malaria to rage in developing countries. Environmental ethics means doing the right thing. But what is the right thing? Should society allow the development of a large manufacturing plant in a wild area near a small town where it could employ thousands of workers but would deprive wildlife of a place to exist and thrive? This industrial use of the land would also deprive hearty hikers, campers, and hunters of a place to participate in their activities. Isn't this an ethical dilemma?

The solutions to this dilemma range from denying a company the right to build the plant to dividing the wilderness area and providing a special cordoned-off section in the wilderness sector to encourage more outdoor participation. The wilderness area could be furnished with campgrounds and comfort facilities to give hikers a place to gather and stay overnight after their wilderness adventures. How about the wildlife in the proposed factory site? Even if the rabbits, squirrels, coyotes, foxes, etc., are captured and relocated, there is still the plant life and millions of small critters to consider.

Environmental agencies have various criteria for dealing with these matters but seldom take into account the specific details of a particular locale. They may provide a forum to allow the opponents and proponents to offer opinions, debate, and arrive at a negotiated compromise. They can define elements such as an endangered species list to be evaluated in the development approval process. The agency could designate one of its officials to sit in judgment and make a decision or delegate the task to a committee of experts to perform in this capacity. If the intensity of the dilemma warrants it, the matter could go to the courts for a ruling.

2.1.1 Anthropocentrism

Anthropocentrism assumes that the environment exists to provide material gratification to humans. In a prior section we reviewed the values derived from various religious faiths as well as atheism, which provide the ethical context of each viewpoint. Many of our modern attitudes regarding nature grew out of the ancient beliefs of the Greeks, East Indians, and Hebrews. Many of these traditions made human beings the exclusive center of the moral universe. The Judeo-Christian Bible, in the story of creation, makes the special place of human beings in the divine plan very clear: "So God created man in his own image, in the image of God created he him; male and female created he them. And God blessed them, and God said unto them, Be fruitful, and multiply, and replenish the earth, and subdue it: and have dominion over the fish of the sea, and over the fowl of the air, and over every living thing that moveth upon the earth."[1]

"People may debate the meaning of 'dominion over,' and those concerned about the environment claim that it should not be regarded as a license to humanity to do as they will with other living things, but rather as a directive to look after them on God's behalf and to be accountable to God for the way in which they are treated."[2] Others counter this view with the example that God, in retribution, drowned nearly every animal on earth in order to punish man for his wickedness.[3] Is it any wonder that some people think that the flooding of a river valley to build a power dam is not worth worrying about? Even after the flood waters had receded, there is a more onerous repetition of the granting of dominion in the Bible: "And the fear of you [man] and the dread of you shall be upon every beast of the earth, and upon every fowl of the air, upon all that moveth upon the earth, and upon all the fishes of the sea; into your hand are they delivered."[4]

When Christianity prevailed in the Roman Empire, Christianity also absorbed ancient Greek attitudes toward the natural world. Father Thomas Aquinas, who wanted to meld Christianity with the Greek philosopher Aristotle, ordered nature on the basis of reason—animals with less reasoning ability exist for the sake of those with more ability. Aristotle believed that "plants are created for the sake of animals," he wrote, "and animals for the sake of men."[5] "Since nature makes nothing purposeless or in vain"[6] (which is interestingly consistent with the Hindu belief in the cosmic order of bosons to multiverse), it is arguable whether nature has made all animals for the sake of man. Vesilind[7] refers to this old quote from Aristotle in her discussion of the anthropocentric view of environmental ethics. Aquinas only recognized sins against God, us, or our neighbors. He saw no possibility of sinning against nonhuman animals or against the natural world.

According to this strain of Judeo-Christian thinking, God gave humans dominion over the natural world and does not care how we treat it. Nature itself is of no intrinsic value, and the destruction of plants and animals cannot be sinful unless by this destruction we harm human beings. The dominant Western tradition today is that the natural world exists for the benefit of human beings. Humans sit atop the food chain and are the only important members of this world.

2.1.2 Biocentrism

Nearly half the people in our world still hold, in keeping with the dominant Western tradition, the notion that the entire nonhuman natural world is only valuable inasmuch as it benefits human beings. But Professor Singer asks whether we should limit ourselves to a human-centered ethic.[8] Biocentrism places the biological world at the center of the planet and focuses on the intrinsic value of life. It does not include usefulness to human beings in its criteria. It questions whether the Western tradition limiting environmental values to service to humans is based on a defensible argument. Many nonhuman animals are also capable of feeling pain, as humans are. These animals can certainly be miserable and may experience some type of joy. Many mammals can suffer due to separation from their family group and even celebrate rejoining the group.

Humans share the world with members of other animal species capable of feeling pain and suffering, of having well-lived or poorly lived lives. One of the three key concepts of Hindu moral conduct urges one to treat all God's creatures with kindness and compassion. Similarly, a key precept of Buddhist moral guidelines is not to destroy life—avoid killing other humans and minimize the suffering of other species. We should regard animals' experiences as having the same kind of value as those of humans. Inflicting pain on nonhuman creatures should be as weighty a misdeed as inflicting pain on human beings. The deaths of nonhuman animals, considered independently from the suffering that often accompanies death, should also have moral importance, though maybe not to the same extent as the death of human beings.

Suppose an area in Montana is the development site for a manufacturing facility. The facility is growing and employing hundreds of workers, and its energy demand is also growing. The local utility proposes to construct a dam on the local river at a site where the river cuts through a canyon with steep sides, an ideal site for an artificial lake. Funding for the project is obtained. The environmental impact statement (EIS) details how this project will benefit the area by creating jobs and generating low-cost electricity without polluting the atmosphere. From the same EIS, a nongovernmental organization (NGO) with a mission to protect the environment observes that the project will also significantly impact the local ecosystem: certain animal species will be lost, and trees that have stood for thousands of years will die. The NGO argues that an entire local ecosystem will be destroyed and that the public should give the preservation of the animals, trees, and overall ecosystem consideration independent of the economic, recreational, or scientific interests of human beings. Here we have a fundamental moral disagreement about what kind of beings ought to be considered in these deliberations.

It might be said that it is an ethical blot on human civilization that it brushes aside the needs of nonhuman animals to satisfy minor human needs. When displaced, a few animals may be able to move into neighboring areas that are suitable, but wildernesses are not full of suitable niches awaiting an occupant. Some species, such as polar bears, are native to certain regions and will only survive in those parts of the natural world. If there is a territory that can sustain a native animal, it is most likely already occupied. In the case of polar bears, most would either drown or starve, neither of which is an easy way to die.

2.1.3 Nature

Living creatures of all types have been considered above, but what about nature itself? Should the intrinsic value of the beauty of surroundings be considered further? Do humans have a moral responsibility for running streams, waterfalls, spectacular canyons, etc.? Some may argue that the destruction of complex ecosystems, the blockage of a wild river, and the loss of rocky gorges are factors to be taken into account only when they adversely affect living creatures. Even unmitigated anthropocentrism does not rule out concern for the preservation of nature as long as that concern can be related to human well-being. Mining uranium or building a nuclear power plant can be challenged if the radiation from those activities can be shown to be a human health hazard. Similarly, many arguments of this type can be made against pollution, connecting the release of gases to human health hazards or to the destruction of the ozone layer, which in turn may be a threat to humanity. Since humans need an environment in which they can thrive, preserving such an environment can be a value within even a human-centered ethical framework.

One argument for preserving wilderness on earth is that it is rapidly disappearing. When a forest is cut down, say, for logging, the continuity of that forest is broken. The forest may regrow over time, but it will never be the same as it was before. The gains made from logging the forest—such as employment, business profit, export earnings, cheaper lumber, and paper—are short-term. Once a forest is cut down or flooded due to damming, the link to its past glory is gone forever. True wilderness is now highly valued because it is scarce, and for this reason environmentalists speak of wilderness as a world heritage. We have inherited wilderness from our ancestors and must preserve it as a world heritage for our descendants if they are to have it at all.

This argument does not imply that there can be no rationalization for cutting down any forest, but from a sustainability standpoint it does mean that any such justification must take the value of the forest for the generations to come fully into account. Opponents to this attitude say that wilderness has little value because it is merely aesthetic, but people go to great lengths to preserve beautiful things—great paintings by Picasso, Gauguin, Rembrandt, Van Gogh, etc. When one walks into a natural setting and pauses to observe the view from a rocky peak or sitting by a stream gurgling over moss-covered boulders, this can be an aesthetic experience, even a spiritual experience to some. When natural settings are destroyed, generations to come are denied of their beauty. Thus, a human-centered ethic can be the basis of powerful arguments for environmental values. From the perspective of such an anthropocentric ethic, economic growth can be seen as something that brings gains to the present and possibly the next couple generations, but only at a price that will be paid by every generation to come. The price to be paid by future human beings may be too high.

2.1.4 Balancing Human Need and the Planet

An environmental professional is often faced with balancing human needs and environmental considerations. Scientists and engineers have the mission of generating energy to meet the heating, electricity, and transportation needs of the industrial, commercial, and residential sectors. Every source of power has impacts on the environment: oil wells, oil shale, and tar sand have adverse impacts on the environment; nuclear power poses a serious threat to health and safety in the event of a structural failure due to an earthquake, tsunami, or equipment malfunction; wind power has an impact on the bird population; hydroelectric plants often involve backing up streams to form lakes that submerge wilderness areas, impacting regional flora and fauna. But the comfortable life to which humans have grown accustomed requires power. As population increases, cities grow, roads multiply, and there is a greater demand for factories to manufacture products that make life more comfortable and processes more efficient. All of this encroaches on nature.

In the 1960s and 1970s, the United States recognized this dilemma and issued the Environmental Protection Act. In order to create new industry or housing sites, etc., proponents must submit an environmental impact statement listing all the benefits, health hazards, and civil and wildlife hardships the project might create. The Environmental Protection Agency checks the accuracy of the statement and evaluates whether the social benefits justify the costs to society. Workshops are conducted for public awareness and to allow impacted parties to raise objections. The process can be arduous and adversarial, and occasionally requires court intervention. The advocates and their supporters, as well as the protestors, make concessions until a satisfactory environmental impact statement is authorized. Often the virtues and ethics of all stakeholders are strained before an agreement is reached, and professional ethics itself is often on trial in very adversarial EIS settlements.

Some would say these EIS agreements prevent rich industrialists from harming the public or wildlife for the benefit of their corporation. Corporations would argue that they will provide jobs and other benefits to the area. The attorneys for both sides are often the beneficiaries. The veracity of expert testifiers is challenged. It can be an ethical battle. In the end, however, the government officials have to accept or reject the project. This becomes a classic battle of choosing among alternatives based on stakeholder impact and using ethical judgments in a public policy context, ranging from small projects to pipelines that run through nations. There will be winners and losers; lobbying and interest-group conflicts, and people and species may be displaced. There may be exploitation of natural resources, but for serving some beneficial goals.

2.2 Environmental Economics

2.2.1 Economic Theories and Public Goods

Environmental economics is an applied field of economics and covers the study of the impact of the economy on the environment and the significance of the environment to the economy. It differs from ecological economics, which takes a biophysical view, and also from resource economics, which deals with the production and use of natural resources. Environmental issues are often tough sociopolitical choices, and economists continue to battle several market myths in these highly charged debates of public opinion. It is worth reviewing how economists are perceived to see environmental issues. Let us explore some myths.

2.2.1.1 Myth of Markets

The myth of the universal market is that an invisible hand of market efficiency solves all problems. Private markets are perfectly efficient with no government interference, if they meet certain conditions. There should be no public goods, no externality, no monopoly buyer or seller, and no unlimited scale of economy. Stock markets meet most conditions. Environmental pollution, however, has a negative externality, and natural resources are public goods. Economists know and understand that universal market does not solve all problems. They recognize that a laissez-faire policy for market externalities (e.g., environmental pollution) leads to social inequities and requires government intervention to ensure fairness and justice for all people.

The myth of market solutions assumes that economists will always recommend market solutions, but this is not true. Economists recognize that tradable permits or emission fees to cover social costs may work for uniform mixing pollutants (e.g., acid rain causing sulfur dioxide from power plants) because they meet most conditions necessary for market efficiency, but the same market-based solution will not work for local impact hazardous air pollutants (e.g., benzene) because that could lead to hot spots at high-impact locations. Rather than control or minimize benzene emissions from the manufacturing facility, it is easier to buy benzene emission credits from a large number of low benzene emission impact locations and satisfy benzene emission reduction credit requirements if that is set up as a government regulatory mandate to get operating permits. This often leads to the environmental regulatory authorities permitting an unusually large number of manufacturing facilities that emit benzene in one concentrated area, creating a neighborhood that has very high levels of benzene in the air, referred to as hot spots for benzene. These in turn result in very high levels of benzene in the air in lower eco-strata locations, for people living closer to those high benzene emission manufacturing plants. This is referred to as environmental injustice, where lower eco-strata people are subject to very high levels of health hazards in the form of benzene in the air.

The myth of market prices blames economists for using market prices to evaluate even nonmarket solutions. Economists realize that the economic value of damages to human health is much higher than the sum of health care cost and lost productivity. They recognize that human suffering in terms of stress and pain of trauma to self or to a loved one far exceeds the obvious tangible economic impacts. Similarly, the economic value of the Amazon forest is much greater than just the aggregate market value of future medicines in yet to be discovered herbs, the value of millions of trees that serve as carbon sinks, and the value of lost ecotourism. The forests have a nonuse value based on just being there, wilderness as a world heritage for future generations of humanity. However, for the purpose of prioritization of resource allocation, aggregation of these factors becomes unavoidable.

Finally, the myth of market efficiency blames economists for being concerned only with market efficiency and ignoring equity or distribution effects. Distributional equity is difficult to value. Environmental regulations are not yet good at resolving this issue. Environmental justice in the United States is a first step, and lawsuits and activism continue to push it. Economists are only now learning to incorporate these equity or distribution effects through multiattribute trade-off analysis.

2.2.1.2 Market Externality

In 1920, Arthur Pigou, a Cambridge University economist, developed a concept of economic externalities. Market exchange generally involves a supply and a demand side, and together in a free market situation, market efficiency is achieved. However, if the seller and buyer market exchange process impacts an external entity that has no say in setting the exchange price, a market externality is created. For instance, a producer continues to produce a product and does not incur the cost of pollution. The producer, given the lower cost of production incurred, can sell the product cheaper and yet make a profit, and the buyer will increasingly consume more of that product at a lower price. However, the community impacted by the pollution from the producer has no say in the pricing process, and that creates a negative externality. Positive externality is where there are public benefits from a market activity and government subsidies are needed to avoid undersupply of a product. The U.S. Medicaid program to provide health care to low economic strata individuals is an example of this. In the environmental context, tax incentives for renewable energy projects are a prime example. Negative externality is where the social cost of market activity is not covered by the economic exchange, and this leads to product overconsumption. A tax equal to the value of the negative externality results in market efficiency. The social cost of pollution is often greater than the private cost to the polluter, and the government needs to intervene with a tax on pollution to minimize the impact on external constituents. These constituents are often outside the market exchange, but get exposed to the pollution generated by the market activity. These are Pigovian taxes or fees, generally paid to the government. Not an emission fee, it is the marginal savings from not eliminating pollution at the optimal pollution level, or "a fee paid by the polluter per unit of pollution exactly equal to the aggregate marginal damage caused by the pollution at efficient level of production."[9] The equi-marginal principle, in controlling emissions from several polluters, requires that the marginal cost of control be equated across polluters to achieve emission reduction at the lowest possible cost. Pigovian fees satisfy the equi-marginal principle, since all polluters set their marginal savings at the same value, that of the Pigovian fees. It is worth noting the difference between effects of fees versus subsidies on market efficiency. In the environmental context, the polluter pays fees or gets tax subsidies for installing pollution control. Are these similar? If not, which is better? Subsidies breed inefficiency. A subsidy may allow firms to continue, while if a company has to pay fees, it may cease to exist. Subsidies make polluting processes seem more attractive. Fees have the opposite effect, promoting innovation or efficiency.

2.2.1.3 Classification of Goods

Classification of goods is a very fundamental concept in economics and quite relevant to the environmental space. Excludability, a basic trait of a product that allows assignment of private property rights, is a necessary condition for the invisible hand of market efficiency to work. A valuable product is excludable if it is practical and feasible to selectively allow consumption of the good (e.g., cars, food, movies). A harmful product is excludable if it is practical and feasible to selectively allow consumers to avoid consumption (e.g., cigarettes). Ethics plays a significant role in making choices between valuable and harmful products; most products offer both benefits and risks. A common example is a medicine that cures and yet has undesirable side effects. Should it be permitted, what levels of constraints are needed to prevent misuse, e.g., a habit-forming prescription drug? Similarly, rivalry is another characteristic of a product where consumption by one individual eliminates the possibility of another to consume or enjoy the same. For instance, beverages, food, clothes, and airline travel are classified as rival products. In these situations, equitable production and distribution have a significant ethical component. However, many public goods, such as defense, public parks, cyberspace, and urban air (both clean and polluted), are not excludable, and this leads to externality and market failure.

2.2.1.4 Tragedy of the Commons

In 1968 Garrett Hardin highlighted the problem with the application of invisible hand of market efficiency-based solutions to common goods.[10] In the well-articulated context of the food basket, he describes an open-to-all grazing pasture where anyone may keep as many cattle as desired. But when its grazing capacity is exceeded—a tragedy—all the cattle die. When, food baskets are created, that is parcels of pasture are assigned private property rights, the grazing is limited to match nature's ability to replenish, and that averts the tragedy. In the context of environmental pollution, instead of taking food from the pasture, air or water pollutants are discharged into the common areas at a rate that is way beyond nature's ability to recycle. But, unlike the food basket, air and water cannot be readily boxed: they are public goods. Assigning private rights to public goods like parcels of air and water, more correctly termed air and water sheds, is not practical, but more importantly, the polluter has no incentive to limit pollution. Thus, putting into practice the private property solution from the context of the food basket tends to favor more pollution. This is market externality, discussed in Section 2.2.1.2, and requires government intervention to ensure clean air and clean water sheds to protect public health. This calls for a redefinition of property rights.

2.2.1.5 Property Rights to Overcome Externality

In 1960, Nobel laureate economist and centenarian Ronald Coase demonstrated theoretically that well-defined property rights could overcome the problem of externality.[11] He demonstrated how enforceable property rights make goods excludable and allow markets to work. He proposed that if stealing is not illegal, nearly all goods are nonexcludable. Any common privately owned property such as a car is excludable and belongs to the owner, only if the owner's right is enforceable. In the context of pollution, one could think of the "right to no pollution" or the "right to compensation." This establishes the "right to pollute as a property."[12] This assigns property rights to an air shed or water shed. If this right to public goods like clean air or clean water could be enforced, these goods become excludable. But is there a right to pollute? Who has the right—the polluter or the victim? Coase showed that if there are no trading barriers, trading of this right to pollute leads to market efficiency regardless of whether the polluter compensates the victim or the victim pays the polluter to not pollute (regardless of how abhorrent the concept is).

Per Coase, although counterintuitive, initial assignment of property rights does not matter for economic efficiency. However, this is true only if it meets several conditions, provided that everyone has perfect information, consumers and producers are price takers, court systems in place for enforcement purposes are costless, producers maximize profits, and consumers maximize utility. Also, there are no transaction costs and there are no income or wealth effects. But if any of these conditions fails, initial assignment of property rights does matter.[13] Transaction costs matter—costs of trading rights must be minimized. If impediments to trading exist, initial distribution matters, and makes a difference to the involved parties. Therefore, vesting pollution rights through allocation of emission credits, for example, is like handing out cash. Also, victims should not be loath to pay, which is a psychological barrier and not an economic issue. While it rarely happens, Santa Maria, California, citizens paid to relocate a cattle feed lot to eliminate pollution. Market efficiency was attained, although the victims paid for it; in that case, the right to no pollution was worth more to the victims.

Finally, markets become efficient with no concern for equity, and this has political significance. Take this political economy model, for example: The government induces polluters to take socially desirable actions, which may not seem to be in the polluters' interest. Government intervention may be in the form of environmental regulations, economic incentives, or trade restrictions. But determining exactly the level of pollution best for society is a complex problem. These are discussed in greater detail in the following section on economic instruments and regulatory policy mechanisms that the government may use for intervention.

2.2.2 Economic Instruments and Regulatory Policy

How much environmental protection is appropriate? What is the right balance between environmental protection and use of natural resources? This is a normative sociopolitical question. Normative economics, however, is value-laden. Individual/social preferences can be biocentric, anthropocentric, or centered on sustainability. Biocentrism has the biological world as the center, focused on the intrinsic value of life, and does not consider usefulness to human beings to be one of its instrumental values. Anthropocentrism assumes that the environment is there to provide material gratification to humans. Sustainability strives to preserve the integrity of ecosystems. The commonly quoted definition comes from the Brundtland Commission, which calls for meeting today's needs but retaining the ability of future generations to meet theirs.[14] Nobel laureate Robert Solow suggests that what we do now must ensure that the next generation lives as well as the current generation. However, he goes on to say that man-made capital (e.g., machines) is a substitute for natural capital.[15]

There are, broadly, four major regulatory mechanisms that governments can utilize to manage environmental issues. These are command and control, liability exposure through lawsuits (also known commonly as toxic torts), emissions fees or tax incentives, and emissions credit trading. Command and control is the most dominant form of regulation, in which specific pollution control equipment, technology, or emission limit for type of plant or specific pollutant is prescribed by regulation. These regulations carry significant financial and personal criminal liability/penalties for noncompliance. They are enforced through the legal framework and courts. The "liability" approach—polluters are responsible for the consequences, the polluters pay for all damages—creates incentives for the polluter to take precaution, as this approach can lead to toxic torts. Some jury awards may be very significant. Because of their ease of implementation, these two, command and control and liability approach, are more commonly used by governments as regulatory mechanisms. However, they do not offer incentives to promote free market efficiency, entrepreneurial ventures, or innovation.

The other two mechanisms offer more direct economic incentives. "Emissions fees" calls for a charge per unit of pollution—it is in the polluter's interest to reduce pollution to lower the fees it must pay. This approach could achieve predefined environmental standards at the lowest possible cost. But the control authorities often do not know the correct fee to charge in order to reach the optimum pollution for market efficiency. Also, this method works better when tradable permit transaction costs are high. "Marketable permits/emissions credits trading," on the other hand, allows polluters and speculators to buy and sell rights to pollute. It separates who pays and who installs controls.[16] Making credits transferable makes them private property and tradable. Trading changes the permit to pollute (a command and control approach) into an economic incentive. A polluter may install excess controls that yield emission reduction credits (ERCs), and use the revenue from the sale of ERCs to pay for the controls. Efficiency is achieved through the use of purchased or internally generated ERCs to avoid controls for operating facilities and equipment units that are more expensive to control. This approach works better for uniform mixing pollutants where no hot spots are created. For example, it has reduced acid rain pollutant compliance costs by $10 billion in the United States. However, spatial intensity and the consequent hot spots, or initial allocation of the right to pollute and the consequent impact on equity, are continuing issues. Per Coase's theorem, both marketable permits and emissions fees could allocate controls in a cost-effective manner to precisely meet predefined pollution levels.[17]

2.2.2.1 Valuation and Cost–Benefit Analysis

Society does not have unlimited resources, so it becomes necessary to allocate resources in an optimal manner. Cost–benefit analysis highlights trade-offs in social investments and answers how much regulation is enough. Environmental and sustainability issues are well-known sociopolitical choices. Society decides if it is willing to pay for the benefits from government policies, and if so, how much. Viscusi argues that, if anything, monetizing these human health risk reductions and environmental protection benefits highlights their economic value and allows policy makers to incorporate them more comprehensively when making more rational policy decisions.[18] Clearly, there are dilemmas involved. For example, should there be differences in the value of human health risk reduction based on age and potential contribution to society. While not perfectly or precisely, in the willingness-to-pay methodology citizens get to make those choices. All natural scenery may not draw the same value. Likewise, allocation of limited society resources between the cleanup of one contaminated site and another groundwater aquifer can only be made through a benefits monetization process. These get complicated, not too dissimilar to other public policy choices. For instance, if the society resource needed to avert one premature death, through regulating a soil fumigant level in the water, exceeded US$2 billion, this would exhaust the society's overall resources: several other possible uses of these resources would have to be stalled![19]

Benefits and costs should be used to evaluate economic efficiency by comparing favorable and unfavorable effects of policies. Decision makers should be allowed appropriate use of cost–benefit analysis when establishing regulatory priority (e.g., where to focus their resources—which pollutants and health risks and emissions from what sectors should be addressed first), and it should be required for all major regulatory decisions. For nearly two decades, the U.S. Office of Management and Budget, per presidential order, has been assessing the costs and benefits of various regulatory initiatives, and this aids in making a reasoned determination of whether the benefits justify the costs.[20] External reviews and stakeholder engagements improve regulatory analysis. The benefits and costs should be quantified where possible but presented with a description of uncertainties. However, decision makers should not be bound by strict cost–benefit tests. In particular, because equity is a noneconomic factor, it is crucial to identify important distributional consequences to ensure that ethical choices are made. This is especially true in such areas as environmental justice—not unduly impacting people of lower socioeconomic strata because they tend to live closer to emissions and discharges from manufacturing facilities that pollute the environment. Also, core assumptions regarding the social discount rate, and the value of reducing risks of premature death and of health improvements are required in the cost–benefit analysis of most environmental policy decisions.

Benefit valuation is often one of the most controversial areas for debate in the public policy arena. Both activists supporting chosen causes and business lobbyists (or advocates, as they would prefer to be called) come from divergent perspectives. As discussed in earlier sections, most environmental goods are public and not marketed as a commodity, making it difficult to determine how society values environmental goods such as clean air. The benefits could be both direct, such as less damage to vegetation by lowering the sulfur dioxide levels in the atmosphere, and indirect, leading to greater mobility toward cleaner regions. The value of the benefit could be either user driven, such as the ability to swim in a clean lake, or simply in the form of the existence of nature's beauty, such as the Grand Canyon, which could be valued for vicarious pride or the sense of stewardship to pass on to the next generation, as discussed in an earlier section. The benefits to society could be assessed through reduction of physical damage or in terms of consumer responses to behavioral surveys (e.g., the contingent valuation method), which seeks out an individual's willingness to pay.[21]

As regards benefit valuations, the most sensitive component is the value of statistical life. Estimating a monetary value of human life is considered by many to be an abomination, an unethical and immoral act. However, valuation of projected reduction in the risk of premature mortality is an evolving area in the community of economic and public policy analysis, and a considerable body of academic literature has been emerging on the subject. Value of statistical life is a measure of society's willingness to pay for statistical reductions in small risks of premature death. It has no application to an identifiable individual or to anything other than small variations in individual risks. It is not intended to suggest that any individual life can be reduced to a mere monetary value. Its sole purpose is to help describe the likely benefit of a regulation. Value of statistical life does not take into account the demographics of the population, in terms of age, income, or health status. The value of statistical life-years extended method does address age differences. However, this approach is politically sensitive and does not account for health status. Society's willingness to pay is different from a human capital approach, which is based on lost earnings. The value placed on mortality risk reduction also depends on the nature of the risk. For instance, premature mortality risks from air pollution are experienced on an involuntary basis and are generally uncompensated, while job-related risks are assumed by individuals who presumably have some choice as to occupation and are compensated for taking a riskier job. The starkest manifestation of this can be seen in the safe exposure standards for toxins in the environment for the general public under U.S. EPA standards versus those for workers within plant premises under U.S. OSHA standards.

Finally, a social discount rate to bring benefits and costs to present value for valid comparisons could pose some unprecedented challenges. For instance, some of the benefits in climate change mitigation result in benefits that may be a generation or even a century away. Almost any rate of discount brings the present value to near zero. Uncertainties in science, political sensitivity, complexity, and inconsistency in the methodology of monetizing benefits, emergence of new materials, and discovery of new adverse health effects are additional compounding factors. Environmental professionals routinely encounter ethical dilemmas, such as how much regulation to impose and how to balance human needs, how to incorporate equity and environmental justice, and how to preserve other living species and the natural world.

2.2.2.2 *Risk Management and Resource Prioritization*[22]

Environmental risk management is the systematic application of policies, procedures, and practices to the tasks of identifying, assessing, and controlling risk in order to protect human life, environment, physical assets, and company reputation in a cost-effective manner.[23] The global financial crisis, as well as the recent environmental disasters, such as the BP Gulf oil spill and the devastating storms Sandy, Katrina, and Allison, commonly attributed to climate change, have brought the importance of risk management to the surface. All organizations, and private, multinational, and state-owned corporations, as well as regions and countries, have a need to understand the risks in their operations and how to best mitigate them. It is also critical to recognize and prioritize significant risks and develop suitable informed mitigation strategies. An enterprise-wide comprehensive risk management approach that addresses all stakeholder impacts and all processes of the organization would result in mitigating the downside risks and gaining from the potential upside risks.

In 2004, the Committee of Sponsoring Organizations of the Treadway Commission published an Enterprise Risk Management (COSO ERM) standard that is linked to Sarbanes–Oxley requirements. The ISO 31000 standard *Risk Management—Principles and Guidelines*, established in 2009, is an international standard that provides a structured approach to implementing enterprise risk management.[24] The reference to enterprises covers both the private and public sectors.

Risk is a function of likelihood and consequence. Risk can be mitigated either by reducing the probability of risk or by reducing the potential impact or shifting the burden of consequence. An effective risk management initiative may affect the probability of likelihood, the severity of consequence, or both. It also leads to better-informed strategic decisions, lower cost of capital, improved image, and enhanced community support. Risk can be initially prioritized using a probability-consequence screening matrix approach. The typical steps are to start with assessing where you are now, then develop a scenario of events, including conditional sequence, and identify risk consequences and probabilities. This is followed by locating them on a simple screening matrix of risk probabilities and consequence severity, typically classified into three or five buckets. Risks are then prioritized into high, medium, or low buckets. For instance, all high-consequence severity events and some high-likelihood and medium-severity events would fall under the high-risk category. Once the risks and opportunities are prioritized, options to mitigate those risks are developed. The mitigation options could vary from lowering the event likelihood probability, the consequence severity, or both.

Risk is the effect of uncertainty on objectives; the effect may be negative or positive. Consequences of a risk materializing may be hazards or opportunities, or there could be increased uncertainty. Risks may be short-term operational risks related to routine activities, medium-term tactical risks associated with projects or mergers and acquisitions, and long-term risks linked to strategy. Risk assessment is the identification of risks followed by their evaluation or ranking. "Risk ranking can be quantitative, semi-quantitative, or qualitative in terms of likelihood of occurrence and potential consequence or impact."[25] The objective of risk management is the assessment of significant risks and the implementation of suitable risk response. The seven Rs and four Ts of the risk management process are

1. Recognition or identification of risks
2. Ranking or evaluation of risks
3. Responding to significant risks, commonly known as the four Ts, which are
   a. Tolerate
   b. Treat
   c. Transfer
   d. Terminate
4. Resourcing controls
5. Reaction planning
6. Reporting and monitoring risk performance
7. Reviewing the risk management framework

The intensity of the risk management effort needs to be proportionate to the level of risk based on size, nature, and complexity, as well as aligned with other activities, embedded in routine operating practices and procedures, comprehensive in scope, and responsive to changes. This will assure stakeholders that risk is managed, governance is met, and the decision process is improved. Risk management must be integrated into the culture of the organization, including mandate, leadership, and commitment of the board. Because of the extensive impact of risk ranking on the allocation of society's resources, ethical behavior in conducting risk management and training to do so takes on a very significant role. For instance, one may debate if one pollutant ranks higher than another, or if any one pollutant pathway deserves greater attention (e.g., mercury or arsenic in water discharges or in stack emissions). It gets increasingly more complex as relevant impacted species, regions of the country, and industry sectors, just to mention a few, are brought into the analysis, and they all compete for allocation of limited resources.

2.2.2.3 Malthusian and Solovian Hybrid

Martin presents an interesting contrast in approaches to environmental public policy using the classic Simon–Ehrlich wager on whether human ingenuity will overcome the barrier of exhausting finite resources: Will mankind's insatiable demands exhaust the planet's finite resources? Or will human ingenuity lead to the more efficient use of existing raw materials and the discovery of new sources of supply?[26] This is a classic battle between those who believe in human capital and creativity to overcome barriers and the naysayers. Consequently, this has great impact on the policy choices made on sociopolitical issues such as sustainability. "This led in 1980 to a bet between a prominent ecologist, Paul Ehrlich, author of *The Population Bomb,* and Julian Simon, an economist at the Cato Institute, a free market think-tank. The two camps were dubbed the Malthusians (after the British economist Thomas Malthus, who forecast that population would outstrip food supply) and the Solovians who bet on Robert Solow, also known as Cornucopian, thanks to their belief in endless abundance."[27]

The Malthusian view is that businesses are the culprits and the cause of this doom. They devour a disproportionate share of the world's nonrenewable resources and produce a disproportionate share of its emissions. So consumers and corporations could do more to reduce the resources they consume, and to curb the appetite for consumption. This view states that they should reduce consumption, reuse and recycle, and show some restraint and responsibility. On the other hand, the Solovian view is that businesses contribute and solve the world's problems and generate innovations that reduce use of resources and pollution. This view states that environmental and social problems can always be resolved through exercise of human ingenuity (e.g., synthetic rubber). Growth has no natural limits because technological innovation is unlimited (e.g., wireless telecommunication). Human capital investments increase rate of return (e.g., you). If Malthusians were right, no progress would have occurred, and humanity should have declined or gone extinct. However, if Solovians were right, then why do we have dangerous levels of CO_2 and an ozone depletion hole? Malthusian risk aversion leads to a cautious measure and tax approach to obtain incremental reductions. Solovians would rather increase rewards for speculative research and decrease allowance (e.g., CO_2) to spur innovation. When reconciling was attempted as a form of compromise, the two approaches caused confusion and dysfunction. The lack of clear direction left the industry guessing, and most corporations chose to wait until later to take any significant action to avoid missteps. A holistic view of all factors determines when one is better than the other.

Malthusian moderation or restraint is an inclusive strategy that needs a sense of urgency and lots of small actions and a broad commitment through regulation, economic incentive, and social or moral pressure. It has several examples of success: the 1975 Corporate Average Fuel Economy (CAFÉ) standards, multipane windows in the 1970s, refillable beer bottles and aluminum cans recycled, and more recently, the Leadership in Energy and Environmental Design (LEED).

Slovanian innovation is driven by promotion and is patently long-term—not rapid—and entails government policy and corporations with large budgets (e.g., pricing for alternative energy). Successful examples of necessity as the mother of invention include the U.S. Army's chlorination of water to make it safe to drink, in 1910,[28] and the low-cost, low-technology solar cooker,[29] which reduces pollution and deforestation. Innovation promotion is indeed an accelerator process accomplished through economic incentives. For instance, innovation of radically disruptive technology needs access to risk capital—to finance speculative R&D—and must generate high and sustained revenues in the future. In the energy and environment space, while the U.S. ethanol policy—linked to highly volatile fuel prices—failed, the German solar energy policy that requires grids to buy solar power at five times the cost of power from nonrenewable sources for a specified number of years has spurred significant enhancement in the solar power yield.

2.3 Environmental Advocacy and Select Mega-Issues

2.3.1 Advocacy and Activism

Author advocate Rachel Carson penned *Silent Spring* (1962) and launched the environmental movement in the 1960s. She drew our attention to nature's beauty by writing, "the more clearly we can focus on the wonders and realities of the universe about us, the less taste we shall have for destruction." She also cautioned why the human being was poised for catastrophic destruction of nature and said, "Only within the moment of time represented by the present century has one species—man—acquired significant power to alter the nature of his world."[30] Around the same time, in the early 1970s, a movement called *Chipko* was started in the Himalayas of India by a group of nearly illiterate poor women, who placed their bodies between trees and contractors' axes to protect them from being cut down, and launched a crusade against deforestation, in the process becoming the environmental movement's first tree huggers.

"Environmental advocacy is presenting information on nature and environmental issues that is decidedly opinionated and encourages its audience to adopt more environmentally sensitive attitudes, often more bio-centric worldviews."[31] The stakeholders that impact or are impacted by the business and activities include employees, customers, shareholders, investors, communities and local governments, nongovernmental organizations, suppliers, their employees, and communities; and the environment in which these products and services are sourced, made, sold, used, or disposed of. Sustainable development brings in future generations that do not have a voice in constructing or directing today's decisions. The actions of environmental stakeholders, invariably shaped by their ethical value system, have grown multifold, with significantly greater participation by a wider set of constituencies, driven by the growth of the Internet, providing real-time access to substantial data globally for scrutiny, many-to-many synchronous interactivity with very few barriers, and the ability to connect communities at unprecedented speed. The impetus for action by organizations comes from the recognition of the potential impact on image and stock prices because of society's growing expectation of responsible action on the part of businesses, including society's demand for ever-increasing transparency on performance. Today it has taken gargantuan proportions and has even become the staple of presidential debates of the most powerful nations in the world.

Environmental activism comprises collecting relevant information, getting the message out, building alliances, going public with the issue, organizing demonstrations, participating in rule making, and if necessary, boycotting brands or exploring other legal recourses. Several significant results have been accomplished through environmental advocacy: Love Canal led to passing Superfund laws for cleanup of contaminated sites in the United States, Chipko resulted in restraining deforestation in the Himalayas, Ozone Hole brought about the Montreal Protocol that eliminated the production of fluorocarbons worldwide, and several others have been instrumental to the birth of major global treaties or local regulations. The Kyoto Protocol to address climate change, the Basel Convention to control cross-border movement of hazardous waste, the Convention on Biodiversity, and the Environmental Justice Conference are some of the better-known examples.

Environmental activism is not limited to passionate individuals or non-governmental agencies. Many of them are driven by political leaders to protect the people in their constituencies. As discussed earlier in Section 2.2.1.5, the political economy model is often the basis for government action to promote underserved areas and prevent undesirable products or irresponsible operations for public health and safety. Financial incentives and regulatory mandates with penalties for noncompliance are typical features of these efforts. Many government programs worldwide encourage the development of renewable energy, while others mandate enhancement of fuel efficiency standards for vehicles. Some government mandates require more stringent emissions control requirements, while others incorporate emissions fees or taxes to encourage reduction. Finally, government agencies that manage financial sector operations, such as the stock exchanges, require publicly listed companies to report their environmental performance as well as disclose their material environmental risks in their operations, in both the near term and long term.

Corporate stewardship for managing reputational, physical, and financial risk and entrepreneurship for exploiting business opportunities are often equally effective in accomplishing the mission of environmental activism and advocacy—the end goal being sustainability to protect and preserve our only planet for future generations. Some classic examples in the United States of stewardship include Walmart's greening of supply chains and 3M's Pollution Prevention Pays campaign. Entrepreneurial examples include

GE's Ecomagination (a $21 billion operation in 2011), Toyota's hybrid cars, and Tesla Motor's electric cars (with $400 million revenue projected in 2012). Another aspect of the corporate role in advocacy through industry groups such as the American Petroleum Institute (API) is to ensure that the regulations are reasonable, workable, and optimize the use of society's resources.

2.3.2 Climate Change, Renewable Energy, and Carbon Share

2.3.2.1 Consensus of Scientists and Global Treaties

In 2004 when coauthor Ramanan had lunch with Nobel laureate Mario Molina, who discovered the root cause of stratospheric ozone depletion, at one point discussions turned to who parallels his discovery in the climate change arena. What surfaced quickly was the name of Nobel laureate Svante Arrhenius; indeed, his paper of 1896 describes how carbon dioxide could affect the temperature of the earth. Recent NOAA[32] data show a strong linkage between the surface temperature of the earth and the carbon dioxide level of the atmosphere. While there are many gases, termed greenhouse gases (GHGs), that contribute to global warming, the gases of most concern, based on their abundance and potential to impact global warming (over 99%), are carbon dioxide (CO_2), methane (CH_4), nitrous oxide (N_2O), and fluorinated gases. GHG emissions are expressed in terms of carbon dioxide equivalents and the term *carbon* has been used to represent that here.

The First World Climate Change Conference was held in Geneva, in 1979, and the first assessment report was produced by the Intergovernmental Panel on Climate Change (IPCC) in 1990. In response, a global treaty, the United Nations Framework Convention on Climate Change (UNFCCC), was adopted in 1992 in New York and went into effect in 1994. UNFCCC signatories met in Kyoto in 1997 and agreed to the Kyoto Protocol reduction targets—European Union (8%), United States (7%), and an average of 5% below 1990 levels; however, the developing nations then did not commit to any GHG emission reduction. IPCC findings in its fourth assessment report, in 2007, received general scientific consensus. That same year it received the Nobel Peace Prize. But meetings since then, in 2007 (Bali), 2009 (Copenhagen), 2010 (Cancun), 2011 (Durban), and 2011 (Doha) have failed to result in any binding agreements.

2.3.2.2 Carbon Share—The Prime Environmental Ethics Dilemma

Scientists believe that a target level of 350 parts per million (ppm) of carbon dioxide in the atmosphere could avert the worst consequences of global warming. We passed the 391 ppm level in 2012, per the U.S. National Oceanic and Atmospheric Administration, and anything over zero net increase is unlikely to get us back below the target of 350 ppm. Carbon is a natural resource, but unlike other dwindling substances, in the case of carbon, we have to constrain its use to contain the generation of carbon dioxide. The potential catastrophic consequences of global warming, both physically and fiscally, compound its significance. Given the near omnipresence of carbon in our life today, restraining the allocation of carbon to competing processes is very complex and creates myriad potential ethical dilemmas. In today's carbon-constrained world, this is a prime area for dilemmas in environmental ethics for academics and practitioners alike.

Ramanan introduces the term *carbon share,* which implies the amount of carbon associated with meeting any human need for products or services. "Carbon-share, given its (negative) market externality, is a socio-political choice where science and society provide direction."[33] Carbon share could become the most contentious distributive justice issue globally. "With different countries likely to undertake different levels of climate-change mitigation, the concern arises that carbon intensive goods or production processes could shift to countries that do not regulate greenhouse gas (GHG) emissions."[34] Issues include what metrics to use: carbon in absolute value or per unit of product or per unit of GDP or per person. This is further compounded when one adds the life cycle carbon footprint, of the entire value/supply chain from mining of raw materials to various stages of production and use and finally disposal; where does one draw the line to measure the metric? Additionally, complexity comes from who should bear the burden, reduce consumption, tolerate an increase in cost of production, or pay more for the same exact functionality.

Finally, the regulatory intervention mechanism within and across countries is a fourth factor. Should this carbon management be driven by government-imposed taxes where choices have to be made on load to be shared by industry sectors, or should a free competitive market cap and trade process be allowed to play a role in achieving market efficiency? However, even here the initial allocation of emission limits will have a short-term impact on select sectors, until the final stable efficient situation is achieved. (See Section 2.2 for an introduction to some basic concepts in environmental economics.)

2.3.2.3 Government Intervention to Contain Global Warming

Ramanan[35] further notes that while there are market drivers like consumer demand for environmentally friendlier products and processes, investor and insurance community concern to manage climate change risk, and corporate stewardship from companies like Walmart and GE, government mandates are inevitable and have begun to mushroom and manifest in many regions in multiple forms in North America. These include the Alberta Specified Gas Emitters Regulation, California's Global Warming Solutions Act of 2006, and the U.S. Environmental Protection Agency's Mandatory Reporting Rule to collect data to inform rule making and the regulation of carbon dioxide under the Clean Air Act. The U.S. Securities Exchange Commission[36] has mandated reporting of material climate risk effective in 2010.

Several countries and regions have climate change response initiatives or plans in place. The European Union has pursued a 20-20-20 strategy to reduce emission (20% below 1990 levels by 2020 while reducing energy demand (20%) through energy efficiency and generating (20%) of its energy from renewable resources). China and India have set goals[37] to reduce their carbon dioxide intensity (emissions per unit of their gross domestic product (GDP)) by 40 to 45% and 20 to 25%, respectively, below the 2005 levels by the year 2020, but that gets neutralized by the growth in their GDP in that same time frame. Governments worldwide are concerned, and debates and actions are on at the local, regional, national, and global levels. They range from innovation promotion to incentives on investments to intervention with regulations.

2.3.2.4 Climate Change, Power Generation, and Renewable Energy

Today, anthropogenic carbon dioxide emissions come almost entirely from fossil fuel combustion for electricity generation and transportation, while most methane emissions have their roots in the agricultural sector. Primary energy consumption[38] could be split into electric power, transportation, industrial, and residential and commercial sectors. The sources of energy are classified broadly into petroleum, natural gas, coal, nuclear, and renewable. It is worth noting that while electrical power at 40% and transportation at 28% cover two-thirds, the other third is consumed by the industry, commercial, and residential sectors combined. The U.S. Energy Information Administration[39] projects that in 2035, non-OECD (Organization for Economic Cooperation and Development) and OECD countries will emit 28.9 and 14.3 billion metric tons of carbon dioxide, respectively. This overview provides global projections on carbon dioxide emissions driven by energy demand and provides some insight into the magnitude of the climate change issue.

Focusing on renewable energy, which includes conventional hydroelectric power, geothermal, wind, biomass, and solar/photovoltaic, currently comprises ~10% of the total. It is worth noting that significant support for solar energy is coming from both academic researchers and industry stewards. Lewis[40] presents a strong case for solar energy (and other academic researchers like Wasielewski[41] concur) emerging to be the lead long-term renewable energy option for meeting the energy needs of the burgeoning human population. The same thought was reinforced by Rogers[42] when he said that, while wind may trump fossil fuels in the near term, solar will emerge as the lead renewable energy option in the final analysis.

The non-OECD countries have a major challenge in balancing the need to bring a vast human population from poverty and meet their growing energy and mobility needs, while ensuring that the very same population sector is not exposed to the vagaries and most severe consequences of climate change. Maintaining economic growth, while driving innovation in clean technology and renewable energy through effective intervention in the form of policies and economic incentives, will continue to top their list of priorities. This is very significant among the environmental ethics issues that the practitioners, planners, and leaders will face in this area. Environmental practitioners will encounter various levels of ethical dilemmas in the context of balancing allocation of primary energy sources among various uses as well as various people. For instance, should one oppose a coal plant with clean coal technology and promote a more expensive windmill, based on reducing carbon dioxide emissions, when that may be competing with the priority of pulling a large population from under poverty into a decent level of prosperity—providing electricity to serve the basic need for clean water and lighting?

2.3.3 Hazardous Wastes and Toxic Chemicals in Products

Chemicals have always been, but more so in recent years, an integral part of our everyday life. But when these chemicals are used inappropriately or disposed of improperly after use, they can have harmful effects on humans, plants, and animals. Even when used properly, many chemicals can still harm human health and the environment. When these hazardous substances are discarded, they become hazardous waste. Hazardous waste is also very often a by-product of a manufacturing process—material left after products are made. Hazardous wastes typically have one or more of the following characteristics: corrosivity, ignitability, reactivity, or toxicity. Unless it is disposed of properly, it can create health risks for people and damage the environment. Environmental professionals have the responsibility to follow and ensure that their clients, their employers, and the regulated community follow the precautions defined by the government and the practices established by professional organizations to protect humans and animals from exposure to toxic substances.

When a toxic substance or hazardous waste is released in the air, water, or on land, it can spread, contaminate more of the environment, expose more people, and pose even greater threats to public health and the environment. For example, when rain falls on soil at a waste site, it can carry hazardous waste deeper into the ground and underlying aquifer. Mercury emitted into the air from a power plant can contaminate the air we breathe. If a very small amount of a toxic or hazardous substance is released, it may become diluted to the point where it will not cause injury. However, a toxic or hazardous substance can cause injury or death to a person, plant, or animal if a large amount is released at one time, if a small amount is released many times at the same place, if the substance does not become diluted, or if the substance is very toxic (for example, arsenic). The effects of coming into contact with a substance depend on how the substance is used and disposed of, who (the very young or old or the sick are more vulnerable) is exposed to it, and the level of the concentration, or dose, of exposure, and how long or how often someone is exposed. Also, exposure may be acute, as in short-time but high-level exposure, or chronic or long-term, because of lower-level but repeated exposure.

Exposure to a hazardous substance could happen through multiple pathways: inhalation, ingestion, or dermal contact. Inhalation occurs from breathing polluted air, hazardous gases, or hazardous liquid vapors; ingestion comes from eating contaminated fish, fruits and vegetables, or meat, or drinking contaminated water; and dermal exposure is caused by absorption by the skin through direct contact. Health effects could range from minor irritation to serious illnesses such as cancer, organ failure, or even death. Some chemicals may affect the reproductive systems resulting in genetic mutations.

Hazardous waste is treated and reused or recycled, or stored safely and disposed of. Typical technologies utilized include lined ponds, incineration, landfills, or underground injection.

More promising methods of pollution prevention focus on minimizing waste, reusing and recycling chemicals, finding less hazardous alternatives, and using innovative treatment technologies. For instance, manufacturing processes used toxic solvents for years before the industry realized their hazard. A production line company making a successful commercial product for years now has to revise the ingredients it uses to make that product, may have to spend substantial effort in modifying its procedure, and occasionally the product quality may suffer. A typical incident was a golf club manufacturer in Southern California. For years it used dichloromethane to wipe the golf clubs before applying an epoxy to attach the grip to the club. The local air pollution control agency recommended that it substitute a nontoxic solvent. It worked well enough, but that solvent had a strong offensive odor that nauseated the assemblers and other workers near the site, who complained. It took several years to resolve this issue.

There are many regulations in place worldwide for managing toxic substances and hazardous waste and preventing their release into the environment. In the United States these span an alphabet soup of regulations that range from those aimed at prevention of new toxic substances entering the environment, such as the Toxic Substances Control Act (TSCA),[43] to those aimed at tracking the release and minimizing the release of toxic substances, such as the Toxic Release Inventory (TRI).[44] Other regulations target the handling, treatment, and safe disposal of toxic substances and hazardous wastes, including the remediation of contaminated sites. These include the Resource Conservation and Recovery Act (RCRA) and Hazardous and Solid Waste Amendments (HSWA),[45] and the Comprehensive Environmental Response, Compensation, and Liability Act (CERCLA), commonly known as Superfund,[46] among several others. CERCLA, in particular, deserves special attention for its construct in "retroactive and several" liability that transformed the environmental insurance industry and has drawn attention from the financial sector for reporting material environmental risks. REACH is the European Community Regulation on chemicals and their safe use and deals with the **R**egistration, **E**valuation, **A**uthorization, and Restriction of **Ch**emical substances. The law entered into force on June 1, 2007.[47]

Its adoption had the purpose of combating the "toxic trade." It was a classic example of reaction to an unethical act. Public resistance to the disposal of hazardous wastes in the better-informed developed world had resulted in what became known as the NIMBY (Not in My Back Yard) syndrome. A dwindling supply of potential sites led to an escalation of disposal costs. Seeking cheaper disposal options for hazardous wastes, operators made a move to export toxic wastes to places with minimal regulations and enforcement mechanisms. The Basel Convention on the Control of Transboundary Movements of Hazardous Wastes and Their Disposal was adopted in response to a public outcry following the discovery of imported deposits of toxic wastes.[48]

Deliberate dumping of hazardous waste to circumvent the rules is, unfortunately, not uncommon. Environmental ethics training for the practitioner can play a huge part in eliminating this problem.

2.3.4 Biodiversity—Aspect of Bioethics

Biological diversity—or *biodiversity*—is a term used to describe the variety of life on earth. "Biodiversity is the degree of variation of life forms within a given species, ecosystem, biome, or an entire planet. Biodiversity is a measure of the health of ecosystems."[49] The International Union for Conservation of Nature (IUCN) plays a central role in biodiversity work. Per IUCN, "it refers to the wide variety of ecosystems and living organisms: animals, plants, their habitats, and their genes. It is the foundation of life on Earth and is crucial for the functioning of ecosystems which provide humans with the products and services without which we could not live. Oxygen, food, fresh water, fertile soil, medicines, shelter, protection from storms and floods, stable climate, and recreation all have their source in nature and in healthy ecosystems."[50] Most people give humans a strong preference over other species. But what may please one person may anger another. Building a dam to provide water to a community may destroy wildlife whose habitats were located under the newly formed lake. Ethicists like Peter Singer value wildlife and wild animals with equal status to humanity.[51]

Biological diversity, comprising plants, animals, and microbes, interlinks land, water, and the atmosphere, both chemically and physically, to enable humans and millions of other species to coexist in an interdependent manner. It is everywhere, both on land and in water, and it includes all organisms, from microscopic bacteria and viruses to more complex plant and animal species. Although many data sources and some tools have been developed, biodiversity remains difficult to track and measure precisely. According to the Millennium Ecosystem Assessment, the total numbers of species on earth range from 5 to 30 million, and only 1.7 to 2 million species have been formally identified.[52] But we do not need precise figures to devise an effective comprehension of where biodiversity is located, how it is changing over space and time, what the drivers responsible for this change are, what the consequences for ecosystem services and human well-being are, and the available options for response.

A common currency measure of biodiversity is *species richness*, which is a count of the number of different species with no consideration for their abundance or level. However, for a more holistic view, it needs to be supplemented by other metrics. For instance, another measure, *species diversity*, takes into account both species diversity and species profusion.[53]

Biodiversity, which comprises a large number of ecosystems and natural processes, is a classic example of market externality, and the services provided by many of these ecosystems are not traded in the commodity market. Because these services have no price tag attached to them, financial markets ignore them. Loss of biodiversity and consequent degradation of associated services could be stemmed, if these values are incorporated in policy decisions. The Economics of Ecosystems and Biodiversity (TEEB) study is working toward this. "The study aims to develop mechanisms to assess the value of nature, drawing attention to the global economic benefits of biodiversity and highlighting the growing costs of its loss."[54] TEEB study claims that while biodiversity is economically valuable, it is not built into private and public policy decisions. Exploding global population and the associated increase in land use and urbanization are the major reasons for loss of biodiversity. A better understanding of local and global benefits as well as the total cost of revival of biodiversity is required to arrest further loss.

The Convention on Biological Diversity (CBD) is a global treaty that entered into force in December 1993. It has three main objectives: conservation of biological diversity, "sustainable use of the components of biological diversity, and the fair and equitable sharing of the benefits arising from the utilization of genetic resources."[55] Other very relevant global treaties to protect biodiversity are the Convention on Wetlands of International Importance (also called the Ramsar Convention), the Convention on International Trade in Endangered Species of Wild Fauna and Flora (CITES), and the Convention on the Conservation of Migratory Species of Wild Animals (also known as CMS or Bonn Convention).[56] Ramsar is an intergovernmental treaty that provides the framework for national action and international cooperation for the conservation and wise use of wetlands and their resources. Note that wetlands hold a significant number of species. The Ramsar Convention is the only global environmental treaty that deals with a particular ecosystem. The treaty was adopted in the Iranian city of Ramsar in 1971, and the convention's member countries cover all geographic regions of the planet. CITES is an international agreement between governments. Its aim is to ensure that international trade in specimens of wild animals and plants does not threaten their survival. Concerned with the conservation of wildlife and habitats on a global scale, CMS aims to conserve terrestrial, aquatic, and avian migratory species throughout their natural habitats.[57]

In the United States, the environmental impact report (EIR) is expected to identify those impacts and try to bring together a compromise that will satisfy the majority of the stakeholders. This can be time-consuming and costly, but the ethic of the environmental professional is to preserve the existing ecosystem to the greatest extent possible, while achieving a balance for the good of all. Clearly this is not always an easy task.

Hundreds of projects are underway around the world aimed at saving species and ecosystems, and providing the knowledge needed for successful conservation action. Biodiversity conservation work is carried out by the International Union for Conservation of Nature (IUCN) through its various programs, including Water and Forests.[58] The IUCN maintains a red list of endangered species.[59] In the pursuit of environmental practice, the professional must consider the impact of actions proposed by industry and government on the ecosystem. Each stakeholder may have a different perspective regarding the impact and value of a proposed change. The job of the environmental professional in conserving biodiversity involves addressing three categories: (1) species and their subpopulations, (2) genetic diversity, and (3) ecosystems. Biodiversity could become a major ethical issue that the environmental profession may be faced with. The environmental profession should then have the objective of developing the ability to resolve ethical biodiversity issues to everyone's satisfaction.

2.3.5 Water Stress

Water stress is a major sustainability issue faced by many countries and regions of the world. In the United States, environmental professionals do not encounter ethical dilemmas around irrigation and drinking water deprivation issues. But even in select regions in the United States, for instance, in Southern California, water supply is a significant issue. We hear of drought areas and locations where runoff is being redirected to supply water to areas of increasing population. It could be compared to a case of stealing from Peter to pay Paul. In making these determinations, the decision maker must consider the ethics of denying one group of citizens in order to benefit another. This has been called hydropolitics.

The state of California is a great example of hydropolitics in action. The area that is currently Southern California, Los Angeles and San Diego, was a dry, almost desert area. Because of the climate and proximity to beaches, this area became highly populated. Northern California had mountains with intense runoff, most of which drained by rivers running through the cities of Sacramento and San Francisco. Sacramento was frequently flooded by this heavy runoff. In 1963, under Governor Pat Brown, construction was undertaken of an aqueduct[60] that bears his name and channels water from Northern to Southern California a distance of approximately 500 miles. Water flows through the Pat Brown Aqueduct in a series of abrupt rises and gradual falls, and after significant falls water is reenergized for use in hydroelectric plants.

A private environmental consulting firm provided comprehensive services to assist the state in complying with National Environmental Policy Act

(NEPA) requirements for the Delta-Mendota Canal and California Aqueduct interconnected projects.

The project was intended to improve the operations and maintenance abilities of the Central Valley to better meet the water supply demands of the region by constructing and operating a connection between the Delta-Mendota Canal and the California Aqueduct. The consulting firm prepared the environmental impact statement (EIS) evaluations, interpreted the results of the study, and helped develop strategies for compliance with the California Environmental Quality Act (CEQA) for the Department of Water Resources and San Luis and Delta-Mendota Water Authority. It also managed coordination with long-term operations, biological opinions, and permit applications. To complete this documentation the consultant had to maintain close coordination with the Department of Water Resources, San Luis and Delta-Mendota Water Authority, and U.S. Department of Interior in order to meet the deadline for a record of decision (ROD) in 2009.

In meeting its obligations, the environmental consultants had to resolve ethical dilemmas almost daily. Regular public hearings were held at which pro and con advocates would present their arguments regarding the issues. Water rights in the northern part of the state and property rights in the path of the aqueduct were at issue. The project team met frequently, often daily, to resolve dilemmas. There were, of course, state-legislated goals and time deadlines to be met that required some skill and discipline in order to be accomplished. Preparation of the environmental impact statement, the report, and the record of decision required comprehensive knowledge of the issues and regulations, as well as excellent debating skills in order to meet the program objectives.[61]

Endnotes

1. King James Bible, Genesis 1:26–28.
2. Peter Singer, *Writings on an Ethical Life* (Manhattan, NY: Eco Press of Harper Collins, 2000), p. 88.
3. King James Bible, Genesis 6:5.
4. King James Bible, Genesis 9:2.
5. Wikipedia, s.v. "Moral Status of Animals in the Ancient World," last modified April 30, 2012, http://en.wikipedia.org/wiki/Moral_status_of_animals_in_the_ancient_world.
6. Peter Singer, *Practical Ethics* (Cambridge: Cambridge University Press, 1993), p. 267.
7. A. Vesilind, S.P. Morgan, and L.G. Heine, *Introduction to Environmental Engineering*, 3rd ed. (Stamford, CT: Cengage Learning, 2009), p. 71.
8. Singer, *Ethical Life*, pp. 93–94.
9. Wikipedia, s.v. "Pigovian Tax," last modified December 11, 2012, http://en.wikipedia.org/wiki/Pigovian_tax.

10. Wikipedia, s.v. "Tragedy of the Commons," accessed December 2012, http://en.wikipedia.org/wiki/Tragedy_of_the_commons.
11. Wikipedia, s.v. "Nature of the Firm," last modified November 30, 2012, http://en.wikipedia.org/wiki/Coase_theorem.
12. William J. Baumol, "On Taxation and the Control of Externalities," *American Economic Review* 62, no. 3 (1972): 307–322.
13. Thomas J. Miceli, *The Economic Approach to Law*, 2nd ed. (Palo Alto, CA: Stanford University Press, 2008), chap. 6, accessed December, http://www.sup.org/economiclaw/?d=Key%20Points&f=Chapter%206.htm.
14. UN World Commission on Environment and Development, *Our Common Future*, Brundtland Report (United Nations, 1987).
15. R.M. Solow, "On the Intergenerational Allocation of Natural Resources," *Scandinavian Journal of Economics* 88, no. 1 (1986): 141–149.
16. P.R. Koutstaal, "Tradeable CO2 Emission Permits in Europe: A Study on the Design and Consequences of a System of Tradeable Permits for Reducing CO2 Emissions in the European Union" (PhD diss., University of Groningen, 1996), p. 17, http://www.unicreditanduniversities.eu/uploads/assets/CEE_BTA/Dora_Fazekas.pdf.
17. Wikipedia, "Nature of the Firm."
18. W. Kip Viscusi, "Monetizing the Benefits of Risk and Environmental Regulation," *Fordham Urban Law Journal* 33, no. 4 (2005): 1003, http://ir.lawnet.fordham.edu/ulj/vol33/iss4/2.
19. W. Kip Viscusi and Ted Gayer, "Safety at Any Price," 1, 2-Dichloropropane in drinking water, p. 58, accessed May 2, 2013, from http://www.cato.org/sites/cato.org/files/serials/files/regulation/2002/10/v25n3-12.pdf.
20. U.S. Office of Management and Budget, *Report to Congress on the Costs and Benefits of Federal Regulation*, EPA Clean Air Act Section 812 Cost Benefit Analysis (1997).
21. Scott J. Callan and Janet M. Thomas, *Environmental Economics and Management* (Cincinnati, OH: Thomson Southwestern, 2010), p. 159.
22. The Association of Insurance and Risk Managers, The Public Risk Management Association, and The Institute of Risk Management, *A Structured Approach to Enterprise Risk Management (ERM) and the Requirements of ISO 31000*, accessed December 2012, http://www.theirm.org/documents/SARM_FINAL.pdf; George L. Head, *Risk Management for Business Executives* (Dallas, TX: The International Risk Management Institute, 2009); Ram Ramanan, "Corporate Carbon Risk Management—A Strategic Framework," *Environmental Manager*, October (2010): 20–23; Ram Ramanan, "Corporate Risk Management Strategies" (presentation at the 15th World Clean Air Congress, International Union of Pollution Prevention and Environmental Protection Association, Vancouver, Canada, September 12–16, 2010); Ram Ramanan, "Outsourcing Divides Owner's Risk for Remediation Projects," *Hydrocarbon Processing* 78, no. 5 (1999): 101–108; Ram Ramanan, "Environmental, Safety and Health Costs and Value Tracking" (presented at the Townley Global Management Center for Environment, Health and Safety; New York: The Conference Board, 1998).
23. Ibid.
24. International Organization for Standardization, *Risk Management—Principles and Guidelines*, ISO 31000:2009 (Geneva: ISO, 2009).
25. Ibid.

26. Roger Marin and Alison Kemper, "Save the Planet—A Tale of Two Strategies," *Harvard Business Review*, April 2012, pp. 49–56.
27. "The Revenge of Malthus," *The Economist*, August 6, 2011, http://www.economist.com/node/21525472?fsrc=rss|fec.
28. Wikipedia, s.v. "Carl Rogers Darnall," last modified September 27, 2012, http://en.wikipedia.org/wiki/Carl_Rogers_Darnall 1910.
29. Barbara Kerr, "Solar Cooker Inventor," accessed December 2012, http://solarcooking.wikia.com/wiki/Barbara_Kerr 1970.
30. Rachel Carson, *Silent Spring* (Boston: Mariner Books, 2002), accessed December 2012, http://www.goodreads.com/work/quotes/880193-silent-spring.
31. Wikipedia, s.v. "Environmental Journalism," last modified December 16, 2012, http://en.wikipedia.org/wiki/Environmental_journalism#Environmental_advocacy.
32. NOAA, "A Paleo Perspective on Global Warming," accessed December 2012, http://www.ncdc.noaa.gov/paleo/globalwarming/paleolast.html.
33. Ram Ramanan, "Corporate Carbon Risk Management—A Strategic Framework," *Environmental Manager*, October 2010, p. 20; Ram Ramanan, "Climate Hot Spots: Analyzing Emerging US GHG Programs" (IHS Forum, San Francisco, CA, September 2007); Ram Ramanan, "A Response to the US Climate Change Debates from a Refinery Perspective," *Hydrocarbon Engineering*, November 2007.
34. Jeffrey Frankel, "Global Environmental Policy and Global Trade Policy—Harvard Project on International Climate Agreements," accessed December 2012, http://belfercenter.ksg.harvard.edu/publication/18647.
35. Ramanan, "Corporate Carbon Risk Management."
36. Securities and Exchange Commission, "Commission Guidance Regarding Disclosure Related to Climate Change," 17 CFR Parts 211, 231, and 241 (release nos. 33-9106, 34-61469, FR-82) (2010).
37. U.S. Energy Information Administration, "Energy Related Carbon Dioxide Emissions," accessed December 2012, http://www.eia.gov/FTPROOT/forecasting/0484%282011%29.pdf, p. 142.
38. U.S. Energy Information Administration, "Primary Energy Consumption by Source and Sector, 2011," accessed December 2012, http://www.eia.gov/totalenergy/data/annual/pdf/sec2_3.pdf, p. 37.
39. U.S. Energy Information Administration, "Energy Related Carbon Dioxide Emissions", accessed December 2012, http://www.eia.gov/FTPROOT/forecasting/0484%282011%29.pdf, p. 139.
40. Nathan S. Lewis and George L. Argyros, "Focus on Next-Generation Solar Energy Technology" (ANSER inaugural meeting held February 12–13, 2008, at Northwestern University in Evanston, IL), accessed December 2012, http://www.anl.gov/articles/symposium-focus-next-generation-solar-energy-technology.
41. Michael R. Wasielewski,* Clare Hamilton Hall Professor of Chemistry, Northwestern University, and Director, Argonne-Northwestern Solar Energy Research (ANSER) Center, and Tobin J. Marks,* Vladimir N. Ipatieff Professor of Catalytic Chemistry and Materials Science and Engineering, Northwestern University (*both PhD advisors of coauthor Ramanan's daughter Charusheela Ramanan), private conversations with coauthor Ramanan.
42. Jim Rogers, CEO Duke Energy (plenary key note speech at the PANIIT Conference on Entrepreneurship and Innovation in the Global Economy, Chicago, October 11, 2009) (also private conversations with coauthor Ramanan).

43. U.S. Environmental Protection Agency, "Summary of the Toxic Substance Control Act," last modified August 23, 2012, http://www.epa.gov/lawsregs/laws/tsca.html.
44. U.S. Environmental Protection Agency, "Toxic Release Inventory," last modified November 29, 2012, http://www.epa.gov/TRI/.
45. U.S. Environmental Protection Agency, "Hazardous Waste Regulations," accessed December 2012, http://www.epa.gov/osw/laws-regs/regs-haz.htm, and FIFRA, http://www.epa.gov/lawsregs/laws/fifra.html.
46. U.S. Environmental Protection Agency, "Cleaning up the Nation's Hazardous Waste Sites," last modified November 16, 2012, http://www.epa.gov/superfund/.
47. European Commission, "REACH," last modified September 14, 2012, http://ec.europa.eu/environment/chemicals/reach/reach_intro.htm.
48. Basel Convention, "Basel Convention on the Control of Transboundary Movements of Hazardous Wastes and Their Disposal," accessed December 2012, http://www.basel.int/.
49. Wikipedia, s.v. "Biodiversity," accessed December 2012, http://en.wikipedia.org/wiki/Biodiversity.
50. International Union for Conservation of Nature, "Improving Knowledge on Biodiversity and Ecosystems," accessed December 2012, http://www.iucn.org/about/work/programmes/business/?6230.
51. Paola Cavalieri and Peter Singer, eds., *The Great Ape Project: Equality beyond Humanity* (New York: St. Martins Griffin, 1993), p. 152.
52. Millennium Ecosystem Assessment, "Ecosystem and Human Well-Being," accessed December 2012, http://www.millenniumassessment.org/documents/document.354.aspx.pdf. Page 29.
53. Wikipedia, s.v. "Species Richness," last modified December 12, 2012, http://en.wikipedia.org/wiki/Species_richness.
54. European Communities, *The Economics of Ecosystems and Biodiversity* (Cambridge: Banson, 2008), accessed December 2012, http://ec.europa.eu/environment/nature/biodiversity/economics/pdf/teeb_report.pdf.
55. Convention on Biological Diversity, December 29, 1993, 1760 U.N.T.S. 79, accessed December 2012, http://www.cbd.int/doc/legal/cbd-en.pdf.
56. Convention on Wetlands of International Importance Especially as Waterfowl Habitat (Ramsar), December 21, 1975, 996 U.N.T.S. 245; Convention on International Trade in Endangered Species of Wild Fauna and Flora, July 1, 1975, 993 U.N.T.S. 243, accessed December 2012, http://www.cites.org/; Convention on the Conservation of Migratory Species of Wild Animals, June 23, 1979, 1459 U.N.T.S. 362, accessed December 2012, http://www.cms.int/.
57. Ibid.
58. International Union for Conservation of Nature, "Business and Biodiversity Programme," last modified February 17, 2009, http://www.iucn.org/about/union/secretariat/offices/rowa/iucnwame_ourwork/business___biodiversity_programme_/.
59. International Union for Conservation of Nature, "The Red List," accessed December 2012, http://www.iucnredlist.org/.
60. Wikipedia, s.v. "California Aqueduct," last modified November 30, 2012, http://en.wikipedia.org/wiki/California_Aqueduct.
61. Ibid.

3

Environmental Ethics and Corporate Governance

3.1 Evolution of Corporate Social Responsibility (CSR)

The role of the public corporation and the nature of the "social contract" have been changing over the past two centuries, but have changed at a faster pace in recent decades. "The corporate scandals and implosions of the past decade, climaxing in the recent global financial crisis and environmental disasters, have highlighted how critical ethically, environmentally, and socially responsible decision making and leadership are to the long-term survival and success of both individual businesses and society."[1] Today, it is impossible to ignore the changing business ambiance and social contract under which corporations have to operate. A principal driver of this societal transformation is the recognition that business is no longer the sole property or interest of a very few. Notably, synchronous interactive connectivity among stakeholders has had a significant role in this change.

Today, with greater recognition that shareholders are only one of many stakeholders, there is a consistent and continual move from shareholder primacy toward stakeholder primacy. Recent CEO survey reports unequivocally demonstrate that sustainability has become a core component of most leading corporate strategies as well as public and private equity interests.[2] A recent study showed that companies committed to environmental, social, and governance (ESG) performance have achieved above-average returns in market capitalization in financial markets. Bloomberg has added ESG metrics to its terminals to serve the rapidly growing group of socially responsible investors.[3]

In just the last two years, several leading articles in the *Harvard Business Review* have focused on sustainability, ethics, and corporate social responsibility. Porter,[4] guru of strategy from Harvard, has introduced the big ideas of "creating shared value" and "higher form of capitalism." Increasingly, leaders recognize that purpose matters and that there is a higher form of capitalism. At the same time, the role of business is transforming from meeting a social contract to realizing tangible economic gains by creating shared value.

Ramanan has effectively captured the impact of these concepts, especially in the context of environmental and ethics education and training.

> In today's global environment, societal needs are defining markets, and business leaders have to address a range of issues from poverty and hunger to sustainability and ethics. Ethical issues include bribery, fraud, green washing, and a culture of corruption. Corporations and leaders have to manage corporate social responsibility and integrate it into their global strategy instead of treating it as just a moral obligation or a risk/reputation management exercise. They also have to endeavor to build new competencies in managing transparency, accountability, stakeholder engagement, ethics culture and social innovation, which are critical for business success in the Next Economy.[5]

Moreover, as the world becomes flat, it draws a large population of poor and untapped human talent into the marketplace, and presents an opportunity to address the bottom of the pyramid. Not only is addressing the bottom of the pyramid the right thing to do, but it is also an opportunity to extract the treasure. Growing disparities in wealth increase risk, while being inclusive may significantly enhance global harmony and growth. Being inclusive is not merely driven by goodness of heart or a pure desire to be good citizens, but has become imperative for long-term survival, and it is a sound business practice. This is also an opportunity for governments and private corporations to partner to serve the people profitably.

Finally, in the environmental context, governments and corporations are facing a plethora of ever-expanding management challenges related to issues spanning physical, financial, geopolitical, and social concerns. Environmental issues are sociopolitical choices, and require "balancing Solovian innovation with Malthusian moderation."[6] Ramanan highlights the need for sustainability education in business schools to train future leaders and decision makers, in what he calls "clarion call for green MBA,"[7] and says:

> With increasing focus on sustainability factors from the marketplace (regulators, investors, financiers and consumers), corporate sustainability reporting is shifting from voluntary to vital. Advances in enterprise systems are making it feasible for corporations to track and transform sustainability performance. The materiality of these seemingly non-economic impacts is the critical link between sustainability and business strategy. Leaders need insight into how to determine which environmental metrics are material to them and relevant to their business.[8]

3.1.1 Advent of Corporate Social Responsibility

The genesis of corporate social responsibility can typically be traced to the occurrence of a visible event with an undesirable social consequence linked to unacceptable corporate behavior. Governments, activists from nongovernment organizations, and the media hold companies accountable for these social consequences. Synchronous connectivity creates rampant awareness and empowerment among the public at large and leads to responses in the form of consumer boycotts, lawsuits, and shareholder resolutions. Communities call for action and governments mandate corporate disclosure of ethical, social, and environmental risks and action to prevent recurrences and mitigate bad social consequences.

The corporate social responsibility debate has intensified in the last decade. Porter lays out a simple and easy-to-read version of the evolution of responses to the question "Why should a corporation be socially responsible?"[9] The early arguments calling it a moral obligation to be a good citizen and do the right thing are not tenable, as they call for individual corporations to make decisions by balancing competing values (e.g., whether to subsidize today's medicine or invest in tomorrow's cure). He goes on to point out the difficulty with the license-to-operate argument as well. While it is obvious that communities do provide permission to the corporation to operate within its confines and utilize its resources in exchange for employment, taxes revenue, products, and services, this approach has serious limitations. Government intervention addressing external elements through a permit process, such as requiring environmental and social impact assessments, has been moderately effective. An inherent part of the process is consulting and engaging stakeholders. However, the process of obtaining stakeholder permission by meeting their needs has a great inherent danger: it could easily lead the corporation and often the relevant community to cede their CSR agenda due to stakeholder and group vehemence. A well-organized and articulate interest group could easily get its way at the expense of what may be more socially equitable.

Building a corporate image and brand loyalty is invaluable in the marketing space. It is recognized as at least a moderately successful technique for promoting the value of corporations' socially responsible behavior. Getting consumer purchasing preference or an insurance to temper public wrath in case of company-caused crisis has been frequently observed and even documented in both academic and industry trade literature. There have been cases of tainted products reaching the consumer market, for instance, Tylenol,[10] but the well-developed public perception of Johnson & Johnson being a responsible company tempered negative reaction from the consumer. While there are numerous anecdotal events to support this kind of tempered public reaction, it still suffers from lack of conclusive correlations to good deeds. These correlations to show that good deeds help temper negative reactions in cases of company-caused crisis are nebulous at best, and there is no conclusive evidence that shows causality. Further, Karnani notes that corporations resist fundamental transformation, and often indulge in cosmetic campaigns and green washing.[11]

Finally, corporate sustainability has been promoted as an act of enlightened self-interest to secure long-term economic performance by avoiding short-term social or environmental cost cutting. A widely quoted definition of sustainable development is "development that meets the needs of the present without compromising the ability of future generations to meet their own needs."[12] A frequently cited interpretation of what sustainability means is "enlightened self-interest that achieves the triple bottom line of often competing, economic, environmental and social goals."[13] Porter criticizes the term as vague and lacking clarity because the required trade-offs remain undefined.[14] Sustainable actions by corporations occur successfully when the triple bottom-line goals are congruent—for example, energy conservation saves money, reduces emissions, and serves more members of society—affordable and available. But there is a huge gap in this model. There is no defined framework to validate trade-offs and optimize value.

Porter suggests that corporations should prioritize the social issues they choose to address based on their relevance to corporations' core business and the value they add, and should only indulge in social issues that are congruent with a competitive advantage for them. Generic social issues—those that are not significantly affected by a company's operations and the resolution of which do not materially affect its long-term competitiveness (such as carbon emission for a bank)—should not be addressed. On the other hand, value chain social impacts are more relevant. They are significantly affected by a company's operations in the course of ordinary business (for example, carbon emissions of the UPS delivery fleet). Further up on the totem pole are the social dimensions of a competitive context—where issues in the external local environment, such as consumer preference or local regulations, may greatly affect a company's current or mid-term competitiveness (for example, carbon emissions from Toyota's automobiles).

Having identified the issues of importance, the implementation of responsive CSR—in which good citizenship mandates efforts to do no harm and mitigate harm that occurs—is a given. This could go beyond legal compliance. For instance, a corporation may apply its more stringent environmental standards to operations in regions with lower requirements. This is common among major corporations in the oil and gas sector as well as many large multinational corporations. This is congruent with their need to protect their image and minimize liability. The transformation from a responsive to a strategic CSR is the effort focused on a win-win, which benefits society while reinforcing strategy—strategic investing in social aspects that enhance competitiveness. Karnani, however, goes on to say that this is clearly a business imperative, and calling it strategic CSR is a misnomer.[15]

3.1.2 Morphing from Basic to Strategic Corporate Social Responsibility[16]

Capitalism in general and the American dream in particular interpret greed to be a healthy trait. "Greed is good—[it] has marked the upward surge of mankind" is the mantra of the early business age.[17] This mantra, along with an obsession with the primacy of shareholder interests, has driven most early entrepreneurial efforts to privatize gains and socialize costs. Greed has become pervasive in business, from executives, corporations, banks, and financial markets. As discussed in Section 1.1 of the book, the speed and scale of these greed-driven actions make them potentially catastrophic. Enron executives, just prior to the failure of the company, took in an average bonus of $50 million while firing employees with a severance of $50,000. Defining U.S. corporations as legal persons mandated to relentlessly pursue shareholder benefits exemplifies corporate greed, especially given that there is no such mandate in U.S. corporate law. The tsunami of toxic assets highlights banking greed at its peak soon after deregulation. Finally, financial derivatives are termed "Wall Street neutrons." They are speculative bets in which most trades are speculations outside real economy. As Visser puts it, "Speculators may do no harm as bubbles on a steady stream of enterprise. But the position is serious when enterprise becomes the bubble on a whirlpool of speculation."[18] The only target stakeholder for a business in the greed stage is the shareholder.

Business slowly morphed in part from pure greed to some form of philanthropy. The realization was that with great wealth comes great responsibility. Business should give back to society to let economic benefits trickle down. But the motto was: first get rich, and then get generous (the power of largess). Over the past century, there have been many iconic charitable leaders. John D. Rockefeller Sr., the founder of Standard Oil, believed in giving back to society, sharing the fruits of success. His approach was called post-wealth generosity—making lots of money first and then dedicating oneself to distribute those riches, leaving a legacy. Andrew Carnegie, the steel magnate, had a three-part dictum: (1) get all possible education one can, (2) make all the money one can, and (3) give it all away for worthwhile causes. Institutional philanthropy—charitable donations directly from business profits to causes like United Way—has been around since the 1900s. Bill and Melinda Gates and Warren Buffett are current icons of business philanthropy. Over the years corporate philanthropy became more professionalized, but philanthrocapitalism—the business effort to do well by doing good—could not yield a superior model of capitalism. President Bill Clinton calls the Clinton Global Initiative a laboratory to test philanthrocapitalism ideas. He says, "The twenty-first century has given people with wealth unprecedented opportunities … to advance public good … our interdependent world is too unequal, unstable and because of climate change, unsustainable."[19] It failed to turn around our global environmental, social, and ethical trends, and it may in fact be distracting us from true systemic sustainability and responsibility.

A more recent phenomenon is venture philanthropy—the idea that corporate foundations can improve effectiveness through monitoring where they invest, providing management support, and staying long enough until those ventures become self-supporting. Other emerging models include traditional foundations practicing high-engagement grant making, organizations funded by high net worth individuals but with all engagements done through professionals, and a partnership model where both the partner and individuals donate the financial capital and engage with the grantees. This stage targets communities as stakeholders through charitable donations.

In the era of marketing, in which perception was promoted as reality and reputation and brand were pushed as the factor that mattered most, CSR arrived as a gift from heaven to public relations. Green washing, which is lobbying against public interest, took on a life of its own. A classic case of ethical failure was the use of smoke and mirrors by the tobacco industry to hide the negative health impacts of smoking by showing macho athletes smoking. Another example is the campaign by contrarian scientists and industry-sponsored think tanks to sow doubt about climate change. Not surprisingly, the same people were involved in both tobacco and climate change issues. Franklin says, "Many companies pretend their sustainability strategy runs deeper.... Need less misdirection and more redirection."[20] At this promotional stage of CSR, public relations is used to reach the public, and in particular consumers and investors.

The management age of business in relation to CSR called for aligning priorities with companies' core operations and competitive strategies. The concept was to embed CSR in management systems, let the voluntary CSR metrics set the bar, and let divisions and the managers make a business case for CSR. This approach made the shareholders and the aggressive NGOs the target stakeholders. A suggested guide was Carroll's weighted CSR pyramid: be profitable, be legal, be ethical, and be generous.[21]

Visser notes that all that happened was CSR moved from peripheral to incremental to uneconomic.[22] However, more of the same CSR will not meet the challenge, and he calls for systemic CSR in this business age of responsibility. At a macro level, CSR shifts from philanthropic, image driven, and standardized to collaborative, reward based, and scalable. It will include innovative partnerships, stakeholder involvement, social entrepreneurship, and real-time integrated reporting. At the micro level, the shift will be from charity projects, product liability, and ethical consumerism to serving the base of the pyramid, social enterprise, ratings, and data streams.

3.1.3 Serving and Seeking Fortune at the Bottom of the Pyramid[23]

Since the 1990s Prahlad and coresearchers at the Ross School of Business at the University of Michigan have argued for corporations to find their fortune at the base of the pyramid (BOP) and eradicate poverty and hunger. There are over four billion people who live under the poverty line but do have need for products and services. Prahlad said, "Low-income markets present a prodigious opportunity for the world's wealthiest companies—to seek their fortunes and bring prosperity to the aspiring poor."[24] Global overcapacity and intense competition in other tiers make this a more attractive opportunity.

The fact that people in the poorest rung of income and wealth are becoming aware of and aspiring to obtain products and services has huge implications for multinational corporations (MNCs). This is a huge untapped market, and MNCs need to create products and services for these populations at prices they can afford.

Strategies for MNCs to make a fortune serving bottom of the pyramid (BOP) include a radically different price performance, that is, new product development, manufacturing, and distribution given the new price structure based on lower purchasing power. Quality has to be viewed in the profoundly different context of robust products for harsh conditions and drastically lower prices. Lower price and lower quality with a price-quality trade-off is an option, but ethical behavior calls for transparency. Some products may not need to last as long. Profitability has to be revisited from the perspective of low margins, very high volume, and high investment intensity. Finally, consistent with the other changes occurring globally, sustainability must be considered—a reduction in resource intensity and use of recyclability and renewable energy.

Karnani counters Prahlad's radical approach that MNCs are to seek fortune at the BOP and eradicate poverty.[25] Karnani says that the poor are vulnerable, lack education, are ill-informed, and often make choices that are not in their interest. The poor get exploited, and he recommends alternate strategies for the BOP. He sides with Nobel Prize-winning economist Amartya Sen's view that "a person's utility preferences are malleable and shaped by one's background and experience, especially so if one has been disadvantaged. We need to look beyond the expressed preferences and focus on people's capabilities to choose the lives they have reason to value."[26] "The BOP approach relies on the invisible hand of free markets to eradicate poverty … instead, the State should extend a very visible hand," said Karnani.[27]

This is a case of market failure and the need for legal, regulatory, and social mechanisms to protect consumers. He recommends that the role of the private sector be to view poor people as producers and create labor-intensive jobs, and not to exploit them as consumers. The growth of private-public partnerships to build infrastructure, especially in underserved markets, is a very visible trend.

3.1.4 Transforming from CSR to Creating Social Impact (CSI)

The previous section presents how CSR morphed from a basic necessity to a strategic need to a systemic imperative. It highlights the link between competitive advantage and corporate social responsibility and shows how the emergence of an integrated systemic responsibility approach drives corporate value. Porter, in a more recent publication, calls for transformation from strategic CSR and redefines the purpose of the corporation as creating shared value, not just profit per se, and presents a new bottom line for corporations to pursue—profits involving a social purpose, one that represents a higher form of capitalism.[28] He defines the creation of shared value concept as "policies and operating practices that enhance the competitiveness of a company while simultaneously advancing the economic and social conditions in the communities in which it operates."

Creating shared value is not simply a matter of philanthropy, sustainability, or executing social responsibility, but is a new way to achieve economic success in which societal needs, not just conventional economics, define markets and social harm is recognized as one that creates internal costs for firms. A basic tenet of creating shared value is that both social and economic progress must use value principles, where value is defined as benefits relative to costs. There are numerous ways in which corporations could create value by reconceiving products, markets, and business models. Examples include offering more nutritious food; harnessing digital intelligence (for example, IBM) to conserve consumer power usage and cut costs (for example, using differential peak rates); reducing use of excess energy, natural resources, or water; producing healthier and eco-friendly products; and delivering books, magazines, software, music, and movies (for example, Netflix) via the web, eliminating materials and reducing costs. In addition, business practices that encourage the ethical creation of value across the value chain could eradicate social harm and simultaneously eliminate costs for the corporation. Examples include not marginalizing small suppliers—eliminating them would hurt company procurement, providing living wages, building supportive industry clusters at the company's locations, and creating open/transparent markets. Employing these strategies provides better incentives for quality and efficiency and ensures the long-term sustainability and reliability of supplies. Creating these positive cycles of economic and social development also leads to seeding new companies and an increased supply of skilled workers.

As discussed in an earlier section, the business ambiance and social contracts have changed significantly in recent decades.[29] Corporations must recognize the strong connections between competitive advantage and social issues. Company productivity and profitability are closely related to environmental impacts (e.g., where the company's operations are dependent on natural resources—supplies may not last if they indulge in unsustainable fishing or logging for timber where nature may not be able to replenish these resources at these excessive rates of depletion), supplier access and viability, employee skills, health and safety, and resource use (e.g., water, energy, rare earths). Porter nicely contrasts CSR and CSV: value gets redefined as creating economic and societal benefits relative to cost rather than doing good. The goal moves from citizenship to philanthropy, sustainability to joint company and community value creation. The driver for action changes from response to external pressure to one that is an integral part of competitiveness. The activity becomes integral in profit maximization. Also, the impact is no longer limited by the corporate footprint and CSR budget, and it realigns the entire company budget. Finally, external reporting and personal preferences do not determine the agenda; the agenda is company specific and internally generated. Governments, with the right kind of regulation, could encourage companies to create shared value, and avoid unhealthy trade-offs between economic and social goals.

The financial sector focused on socially responsible investment is growing fast. It has grown from $2.7 trillion in 2007 to $3.0 trillion in 2010. Investors in this sector only desire to invest in corporations that have been vetted by and are high on the Dow Jones Sustainability Index or FTSE4Good Index.[30] We recommend moving one more step forward from creating shared value (CSV) to creating social impact (CSI). There is a new genre of vibrant young leaders—social entrepreneurs driven by a passion for social cause and keen on making a net impact—they want to do good while doing well.

3.2 Framework for Reasoning to Make Ethical Decisions

3.2.1 Unethical Behavior—Why Does It Occur in Organizations?

Sims offers a list of 12 unethical encounters commonly seen in corporations based on 60 *Wall Street Journal* articles.[31] These are stealing, lying, fraud, influence buying, hiding or failing to protect data, cheating, working below par, personal or organizational abuse, not complying with legal requirements or noticing an act of noncompliance and not reporting it, and arbitrarily choosing one option in case of an ethical dilemma. The list below provides a context for each of these common unethical situations a professional or manager may face.

1. Stealing—raiding another consultant's expert employee to win a consulting job
2. Lying—misrepresenting statements to stay under the applicability of a more stringent regulatory requirement or pollution control equipment
3. Fraud/deceit—misleading consumers using green-washing commercials
4. Influence buying—paying bribes to expedite the clearance of environmental permits
5. Hiding or failing to protect data—for example, hiding the effects of tobacco on personal health
6. Cheating—taking unfair advantage of a less experienced permit engineer or overselling expertise to win a contract
7. Below-par work—providing an incomplete or subpar report so that it meets budget or cost limits
8. Interpersonal abuse—claim credit for a junior's work

9. Organizational abuse—delay the issuance of a permit on a minor technicality
10. Noncompliance—using creative interpretations to avoid a regulatory requirement
11. Accessory—failing to blow the whistle when a client refuses to stop toxic releases
12. Dilemmas—your client's attorney directs you to hold back on reporting the toxic release to the agency under the confidentiality clause of your consulting contract (see Section 4.2.2)

3.2.2 Ethical Decision Making—Characteristics

Ethical decision making is affected by three qualities: competence in identifying issues and evaluating consequences, self-confidence in seeking different opinions and deciding what is right, and willingness to make decisions when the issue has no clear solution. The development of these qualities in individuals depends on their intrinsic personality and their stage of moral development at the point of decision. People could be fatalistic or believe that it is choice and not chance that leads to consequences. Another significant personality trait is Machiavellianism—the willingness to do whatever it takes to achieve a goal. People with this trait are comfortable manipulating others and, unlike Gandhi, believe their desired end justifies the means. Finally, people's cognitive stage of moral development is affected by their group interactions and personal growth. The early self-stage is shaped by reward and punishment. The focus in this stage is on the expectation of others in authority, including parents, peers, and society. At the mature level, people begin to recognize and accept diversity in value systems, and become confident and often fixed in their own choices. The social contract practiced by society defines right and wrong based on principles of justice and fairness. Often individuals also become fixated on their self-selected absolutes for what is ethical or moral.

Other factors that contribute to individuals' ethical decision-making process in the workplace are the moral intensity of the consequence of the action; the individual's empathy, knowledge, and intellectual and emotional ability enabling one to recognize the potential impacts on stakeholders; and the influence of the business and job environment. For instance, sending a personal email from an office computer may not be seen as unethical at all. Some individuals are more capable of understanding the broader impacts of an issue than others. For example, the natural attenuation remediation of a contaminated site may be better understood by someone with expertise in the area. They may also realize that this remediation method is a good use of the community's resources, and that it reduces health risks. However, this does not imply that such an individual is more ethical and will make more ethical decisions. For instance, one could be an environmental activist with a set agenda to make the company responsible for the contamination pay more for the remediation, or one may be an environmental consultant who provides remediation contract work, and choosing this natural attenuation option may eliminate contract work, and hence may lead the consultant to choose the less ethical option of dig, move, and treat the dirt that provides him additional contract work and compensation.

The ethical climate of the organization puts circumstantial or situational pressure on the individual and impacts the decision-making process. More often than not, individuals receive conflicting and confusing messages about ethics. An example of the discrepancy between social norms and organizational practice: being open and honest at home and church conflicts with being secretive and deceitful at work. For instance, when making a competing bid for an environmental consulting service, an effort is made to mislead the competitor, or underbidding is practiced to win the contract. At school and on the football field, the individual is repeatedly advised to follow the rules. This is in stark contrast to a "do whatever it takes and get the job done" mentality from the Machiavelli manager who wants to move up fast. Lessons from home economics and the value of staying within budget are thrown out the window in consulting projects when one hears, in corporations and even universities, that one should use the money before the end of the budget year or else lose it. Teachers tell you to take responsibility and volunteer to serve the community and work even harder, but at work you routinely observe your supervisors passing the buck to you for work by enticing you to be a team player, and once the task is accomplished, they take credit for it and grandstand. Consistently seeing such individuals succeed and move up clearly motivates you to do the same.

Sims draws upon Reidenbach's moral development pyramid and finds similarity between the moral maturity of organizations and individuals.[32] Interestingly, Caroll's CSR pyramid is quite similar and has broadly equivalent criteria.[33] The ethical culture matures from an amoral focus on profitability and an obsession with shareholder primacy to following the letter of the law, in which the understanding is that if it is legal, it is ethical. The next level is termed responsive, and consists of seeking social responsibility as long as it is in the corporation's best interest to do so. This is consistent with Porter's strategic CSR.[34] The next two stages up are called the emerging ethical and developed ethical, in which a balance between profits and ethics along with commitment to codes and ethical principles is in play. This more closely parallels Porter's CSV and Visser's systemic CSR.[35] Regardless of the stage, corporate culture plays a role, and Sims provides a set of nine broad categories that have differing levels of impact on the ethical decision-making process carried out by individuals in the organization. These are driven by ego, benevolence, and principles. Personal self-interest, company profit, and operating efficiency are ego-driven climates. A climate of benevolence promotes individual friendships, team interests, and social responsibility. An organizational ethical climate, at the top of the pyramid based on principles, leads to personal morality/ethics, rules, standard procedures, laws, and professional codes.

3.2.3 Groupthink—A Dysfunctional Phenomenon

"Groupthink is a psychological phenomenon that occurs within groups of people, in which the desire for harmony in a decision making group overrides a realistic appraisal of alternatives."[36] This is an enormously powerful force in today's world of global harmony. In the current context of the ethical decision-making process in organizations, especially corporations, its role can hardly be overplayed. While a normative definition of morality is a set of absolute values, the descriptive definition refers to the values that a group of people at a given point of time thinks are right. The descriptive definition below is particularly relevant to today's organizations and to further discussion of problems with groupthink. "Moral systems are interlocking sets of values, virtues, norms, practices, identities, institutions, technologies, and evolved psychological mechanisms that work together to suppress or regulate self-interest and make cooperative societies possible."[37]

Haidt captures the problems with groupthink brilliantly with the idea of morality binds and blinds.[38] Humans are strategically altruistic: 90 percent chimps and 10 percent selfless and team-spirited bees. However, when taking one for the team, assimilation and acceptance in groups are predicated on getting along and not rocking the boat; dissent from the group is discouraged. This possibly explains the polarization of the two-party political system in the United States. Each group thinks it owns moral superiority. Democrats broadly wrap three of the six pillars of character—justice and fairness, caring, and civic virtues and citizenship—into compassion for the underprivileged. Republicans, on the other hand, loosely encompass the other three pillars—trustworthiness, respect, and responsibility—into the concepts of freedom and the righteous behavior of the individual.

Groupthink is described as a dysfunctional process by Janis, and Sims provides a set of valuable symptoms to look for and avoid: to separate the ethical decision process from the malaise of groupthink.[39] The symptoms include a range of illusions, group morality, unanimity, and a false sense of invulnerability that lead to no reexamination of ethics. Because of peer pressure, which rejects dissent and promotes a selective bias against new information, members of a group indulge in rationalization and self-censorship, and leaders actively shield groups from any external negative feedback. This provides such a false sense of assurance that the correct path has been selected that no one investigates impact on stakeholders who may have been completely overlooked or the possibility of a better solution.

3.2.4 Evolution of a Framework for Making Ethical Decisions

3.2.4.1 Early Foray—The Four-Way Test[40] (1932)

Religions, as discussed elsewhere in this book, have been pursuing the absolute in morals and ethics. Philosopher Socrates called for first learning what is good before pursuing it. Plato assigned cardinal virtues to man by class: prudence or wisdom to the ruler, courage or fortitude to the warrior, temperance or moderation primarily to the worker, and justice to all regardless of class.[41]

One of the early forays into formal ethical decision making by business associations was created in 1932 and adopted 11 years later by Rotary International as its code of ethics. It was both simple and elegant, and has survived the test of time. Every decision becomes ethically robust as it goes through the following four-way test:

1. Is it the truth?
2. Is it fair to all concerned?
3. Will it build goodwill?
4. Will it be beneficial to all concerned?

3.2.4.2 Examining the Ethics of a Business Decision (1981)

Nash differentiates between the ethical expectations from man and the corporate ethos.[42] She places corporate ethos halfway between amoral and Platonic virtues; it fulfills the social contract that centers on avoiding social harm. She sets out a set of 12 very pragmatic questions for ethical inquiry that are easily comprehensible to the business executive. We have added a sustainability context to each of the dozen questions.

1. Have you defined the problem accurately? The focus should be on factual neutrality and not an attempt to colorfully articulate group or individual morality. An example is presenting select environmental data to advocate stringent regulation.
2. How would you define the problem if you stood on the other side of the fence? Calling for articulating the other side (stakeholders) interrupts self-interest and brings in empathy. For a long time, the auto industry resisted any efforts to impose fuel efficiency. However, once they recognized the younger consumer group's call for greener cars, enlightened self-interest kicked in.
3. How did the situation occur in the first place? Investigating history often provides clues to the root cause rather than just the symptoms. Unethical acts often start inadvertently and are not recognized until they become a crisis. Recurring patterns of seemingly trivial events often point to the real issue. This has been most recently seen in the

disastrous Gulf oil spill, where the culture of relentless cost cutting at the expense of plant safety was ultimately discovered as the root cause.

4. To whom and to what do you give your loyalty as a person and a member of the corporation? Conflicts of loyalty are real, and identifying them is just the beginning. All structural precautions, whether they are mandates or industry or corporate standards, are inadequate unless people involved also have a strong enough sense of integrity to put upholding the principles above personal gain. Illegal dumping of hazardous waste to qualify for a personal bonus that is predicated on cost savings is a classic example. This issue in itself highlights the need for structured effective ethical training and a reinforcement of the ethical culture. It also brings to the fore the value of participatory training workshops for staying on top of these situations should they arise, and not being caught unprepared.
5. What is your intention in making this decision? A good intent is not enough to justify making a decision in an unknown area that could result in more harm than good. An oil company learned the hard way when human interaction with the natives of an ultra-pristine area inadvertently led to an avoidable epidemic of the common cold virus, to which the local population had not developed any immunity. The intention was noble—to engage the native stakeholders directly. However, they did react responsibly and brought in all the emergency medical personnel and medication to address and resolve the epidemic.
6. How does this intention compare with the probable result? This question brings into focus two inherent dangers in the decision process—knowledge of the future is often impossible, and blinding overconfidence driven by good intent is a recipe for disaster. Banning DDT is an example—while it may have helped eliminate the human health impact of this chemical, it is claimed to have set off resurgence of malaria and several deaths in tropical regions like India because of the lack of alternate ways to contain the spread of mosquitoes.
7. Whom could your decision or action injure? This is an extremely significant question in today's context. The innovation and new products have significantly outpaced our ability to manage their production and contain their distribution or disposal. Businesses often only investigate potential impacts when hit by lawsuits—as in the case of asbestos. Nanotechnology, genetics, and biological and viral antidotes are prime examples of the unknown.
8. Can you discuss the problem with the affected parties before you make your decision? This is an avant-garde effort. As we are aware today, initiatives such as Equator Principles and UN Global Compact

mandate disclosure and the engagement of relevant stakeholders in the environmental and social impact assessment process prior to funding or permitting a major project.[43]

9. Are you confident your position, though it now seems valid, will continue to be valid over a long period of time? While budgeting and executive compensation systems and, more importantly, speculator investor incentive systems are typically for the next quarter and, at best, one year, long-range planning calls for a much longer time frame. Over time, uncertainty in a business climate is inevitable, more so today when the globe is shrinking and manufacturing and money move at lightning speed. This is an ongoing debate discussed more extensively elsewhere in this book, and it becomes very significant in the case of addressing climate change—an issue that brings in an intergenerational time frame.
10. Could you disclose your decision to all stakeholders as a whole without qualms? This is an absolute necessity in today's world of synchronous interactive connectivity. Today, one does not have to wait for press coverage. In addition, integrated reporting of nonfinancial metrics (e.g., social, environmental, and governance metrics), along with financial factors, is becoming a global norm. As Warren Buffett once said, it takes 20 years to build a reputation, and in today's context about 20 minutes to lose it—and only then do you discover you had a reputation to begin with. CSR reporting and disclosure of material risks in operations is increasingly becoming a mandate in most advanced economies.
11. What is the symbolic potential of your action if understood or misunderstood? The role of corporations in society is becoming increasingly symbiotic. As seen in the earlier sections, CSR has morphed and transformed into a new genre, in which making a social impact is becoming integral to the core business. This significantly increases the visibility of any corporate initiative in the social arena. Charity, philanthropy, or even the creation of shared value through public-private partnerships is subjected to intense scrutiny. Also, perception is reality and matters, so every action has to be evaluated in light of the perception it may create. The recent furor over donations by Chick-fil-A to a religious organization set off protests by the gay and lesbian alliance and impacted their expansion plans.
12. Under what conditions would you allow exceptions to your stand? This question sets the level of consistency with which the policy will be implemented. The waiver of conflict of interest ethical code became a norm in Enron and eventually led to its demise.

3.2.4.3 Ethical Values Checklist for an Environmental Professional (1998)

Taback provides a practical values checklist that gives environmental professionals an opportunity to examine a situation from all sides.[44] The format could be used whenever a professional is faced with an environmental dilemma. It is based on the six pillars of character discussed earlier in Section 1.2.3. It suggests that decision makers create a table that lists the various subelements of the six pillars of character as values on one side, and evaluates the issues and related actions on the other side. The major values elements, as presented earlier, are trustworthiness, respect, responsibility, justice and fairness, caring and civic virtue, and citizenship. The coauthor suggests that after completing such a table, the professional should take the contemplations to colleagues, peers, and supervisors for validation and select the right action.

We present an actual case and its value analysis using this approach in the next chapter of this book.

3.2.4.4 Ethical Principles for Environmental Professionals (2005)

Paine selects eight business-specific ethical principles.[45] These are presented here with expanded sustainability/environmental ethics examples:

1. Fiduciary: The materiality of potential climate change risks, conducting environmental audits of current operations, conducting due diligence audits of companies to be acquired, and setting environmental standards for operations in countries with less stringent standards are all areas to which the fiduciary duty principle applies extensively in environmental professions.
2. Dignity: Protecting public health and safety is one of the primary values of an environmental professional or issue manager. Special concern for the vulnerable—setting air and water quality standards that protect the most vulnerable segment of the public, including children, sick people, and elderly people—is a given. Dignity often goes beyond public health and welfare to include nonhuman species as well.
3. Property: Protecting others' property, including nature's beauty, is inherent in all the decisions that an environmental professional makes. Responsible use of one's property is most evident through minimizing the impact on the natural environmental and leaving the site after its use in at least as pristine a condition as before. It also includes the natural inanimate world as something to be protected for generations to come and enjoy.
4. Transparency: Disclosure of material information is built in to every report or document to which the environmental professional is connected. For instance, any findings on the potential health impacts

of a new chemical or a newly discovered impact need to be documented and reported to the appropriate authorities so they can take necessary protective action. Many infractions of regulatory requirements are managed through self-reporting by the corporation and its environmental professionals.

5. Reliability: The society depends on the reliable performance of the environmental professional. In light of this, one must maintain skills and expertise, but also not make commitments or provide services in areas that are beyond his or her expertise. For instance, a board-certified air quality specialist should not commit to work in wastewater or hazardous waste management areas. Also, professionals should make sure they are not overcommitting, which could result in below-par work and also result in injury to society.
6. Fairness: This principle applies to every aspect of the work the environmental professional performs, especially in dealing with clients, colleagues, and competitors. Fairness becomes particularly relevant when decisions related to environmental policy come up. Environmental justice, in which the lower economic strata are disproportionately impacted by lower environmental quality, is a special case of applying the fairness principle. The central idea of the Brundtland Commission's definition of sustainable development is that of intergenerational equity. Climate change is another classic case of the same concept.
7. Citizenship: Respect for law and regulation is pervasive in an environmental professional's job. Given the market externality of environmental issues, government intervention becomes mandatory. Environmental professionals, given their expertise in an area that could provide public benefit, must participate in advocacy, and many do, especially in areas in which they have expertise relevant to their community.
8. Responsiveness: Addressing public complaints, participating in public hearings and responding to any environmental crisis in which they could have an effective role are all relevant to this principle.

It may be noted from the examples above that each ethical principle could have implications for the moral duties of the environmental professional. Sucher suggests evaluating each action from multiple ethical perspectives.[46] Decision makers have to ensure comfort with the consequences, respect for duties and rights of others, and compliance with community norms and company commitments.

3.3 Corporate Governance and Integrated Reporting

3.3.1 Corporate Governance and the New Social Contract

3.3.1.1 Recession with a Distinction[47]

The Sarbanes–Oxley Act of 2002 was the U.S. Congress's response to the huge accounting scandals of the time, including Enron and WorldCom. However, what surfaced in September 2008 was unprecedented. Several "too big to fail" financial institutions like AIG, Fannie Mae, Freddie Mac, Lehman Brothers, Merrill Lynch, and Washington Mutual fell like dominoes, some of the largest failures in U.S. history. This was many times faster, and much more obscure and global in reach, than any other, including the Great Depression of 1929.[48] Once again, a parallel from an environmental perspective could be drawn from the famous Rachel Carson quotation, "Only within the moment of time represented by the present century has one species—man—acquired significant power to alter the nature of his world,"[49] especially when seen in the context of climate change. Transparency is the only known antidote to complexity. Opacity is the arsenal of some in the financial sectors. Once again, a parallel could be drawn of unethical companies that would rather not disclose their material environmental risks and liabilities, to those that would hide their large financial misappropriations and losses, both for fear of the impact on the next-quarter stock price.

3.3.1.2 Concomitant Integrity Recession[50]

The recession brought job cuts, and with it came employee disengagement. The impact on corporate integrity and ethical culture was immediate. The Compliance and Ethics Leadership Council observed a 20 percent increase in misconduct from the first to the second half of 2008.[51] The survey also found that companies with weak ethics cultures had five times as many incidents of misconduct as those with strong ethics cultures. The council's research also showed very inconsistent treatment of reported incidents, clearly attributable to lack of ethics training.

When employees came to believe that management would respond appropriately to unethical behavior if it occurred, misconduct dropped.[52] This calls for ethics training to equip managers to deal decisively and effectively. As with any other corporate policy, successful implementation requires that the entire workforce, in particular the person who reported the incident, sees how these real-life situations are addressed. In good times, these studies are critical, and in bad times they are mandatory. The business leaders of today are challenged now more than ever by the global economic crisis and impact of the economic climate on corporate culture, which highlights the need for collaboration on issues of compliance and ethics.[53]

3.3.1.3 The New Social Contract[54]

Business used to be charged with producing a functioning product and selling it at a profit, often at the head of a linear chain. Today business is one thread in the complex web of interwoven society. Today business often becomes responsible, cradle to grave, for not just its inanimate inputs and outputs, but for all related human and environmental interactions. Is it safe throughout its life cycle? Does the process violate any ethical principles on its path? Does it meet the most pressing social needs, and if so, does it do so effectively and responsibly? The process brings in the issue of equity, which was for too long a missing component of market efficiency. Water is no longer just a beverage; the question today extends to whether the farmer was equitably compensated for the diverted water: Was it a fair trade? Did it avoid increasing stress on community access to water? An attitude of "that's not my job" will not work. Big companies can no longer outsource negative societal consequences. Leaders will increasingly be assessed on long-term impact on societal well-being.[55]

3.3.1.4 Expanded Fiduciary Duty

Because of the social expectations associated with the move from shareholder to stakeholder primacy, the financial sector is under pressure to commit to sustainability, which calls for measuring environmental and social impacts, continuous improvement of portfolios, proactively fostering sustainability, building capacity, and linking performance—all driven by voluntary measures such as the Collevecchio Declaration, Equator Principles, and OECD Principles. The U.S. financial reform holds banks responsible for their actions long past the date of transaction. Lenders and institutional investors are increasingly required to disclose, through integrated reporting, how their investments are channeled into responsible operations from the perspectives of longevity, risk, and reward. Also, the insurance and reinsurance sector has suffered and taken very substantial steps in dealing with environmental callousness. Swiss Re and Marsh offer carbon risk insurance products. However, they have cautioned clients that senior executives could lose insurance protection against climate change-related liability claims by shareholders if they fail to adopt adequate policies. The U.S. Securities and Exchange Commission (SEC) and several stock exchanges across the developed world call for reporting material risks in their operations as part of their annual financial reports. Investment banks are required to conduct due diligence for material risks, including environmental liabilities prior to preparing prospectus for any new initial public offering (IPO). Today countries (e.g., China) require environmental audits by government-authorized agents before a company is allowed to list its stocks. The European Union is on its way to mandating environmental, social, and governance metrics disclosure from publicly listed companies. These developments clearly define the direction of the new metrics by which corporate leaders are increasingly likely to be measured, and which will potentially determine their compensation in years to come. They include, without a doubt, a call for ethical leadership.

3.3.1.5 Fiduciary Role—Environmental Responsibility

The financial sector, as the economic engine of all businesses, has a very unique role in environmental assurance and risk management. It has the equivalent of a fiduciary duty to channel investments into operations that are responsible from longevity, risk, and reward perspectives. The World Bank and its project financing arm, International Finance Corporation (IFC), have been sued by impacted parties and NGOs for not ensuring that their borrower operates their project responsibly. This lawsuit has led to the development of the industry group voluntary initiative called Equator Principles,[56] designed to manage environmental and social risk in project financing. Although it is led by IFC, signatories include Goldman Sachs and Citigroup.

The precursor to Equator Principles was the Collevecchio Declaration on Financial Institutions in 2002, a move by over 100 NGOs to advocate environmentally responsible behavior in the financial sector.[57] Driven by the social expectations of the move from shareholder to stakeholder primacy, the financial services sector must commit to principles and steps. The first principle is sustainability, calling for measurements of environmental and social impacts, continuous improvement of portfolios, and proactive fostering of sustainability, building capacity, and performance. The second principle is to do no harm, which requires the creation of sustainability procedures and the adoption of international standards. The next three principles involve taking full responsibility for impacts, accountability for public consultation and stakeholder rights, and transparency through corporate sustainability reporting and information disclosure. The final principle is sustainable markets/governance, which covers the following public policies and regulations: recognizing government's role and discouraging unethical use of tax havens and currency speculation.

Equator Principles (2003) relevant to environmental ethics include social and environmental impact assessments (SEAs), compliance with all applicable social and environmental standards, covenants in financial documentation, public consultation and disclosure, grievance mechanisms, independent reviews, monitoring, and reporting. Furthermore, the public consultation and disclosure process requires that all stakeholders be consulted for the SEA, disclosure of the SEA results to the public, and ongoing consultations during construction and operation. These must be conducted in local languages, showing respect for local traditions and ensuring that the groups consulted are representative.

Equity investors are equally concerned about environmental risks and the longevity of corporations. Corporate shareholder resolutions are often filed by pension funds and institutional investors seeking data on companies' risks and initiatives related to climate change, for example, policy, emission reports, and mitigation plans. SEC, to protect equity investors investing in publicly listed stocks, has guided listed companies to manage climate risk like any other business risk. They suggest that the climate risk mitigation may require internal capacity building and stakeholder and community engagement, and warn that uncertainty is not a reason for inaction.

3.3.2 Integrated Financial and Sustainability Reporting[58]

The huge role environmental professionals play in integrated financial and sustainability reporting is related to the materiality and integrity of evaluating and reporting these highly technical and often complicated arrays of environmental factors, and ethical dilemmas are inevitable. Good comprehension of this emerging area and in the context of ethics is crucial for environmental leaders and professionals.

3.3.2.1 Environmental and Financial Factor Interrelationships

An area of considerable debate is whether environmental expenses are costs with no tangible returns. Does investment in environmental performance pay off in better economic performance? According to the World Business Council for Sustainable Development (WBCSD), companies benefit from sustainability by driving down costs through operational performance and efficiency, minimizing business risks, attracting and retaining talent, and increasing employee morale and productivity.[59] Sustainability also generates growth by institutionalizing learning and innovation, creating brand recognition and reputation, building customer loyalty, improving access to capital, improving supply chain management, and providing a long-term strategic focus. Sustainability brings clear tangible benefits.[60] For instance, companies in the Dow Jones Sustainability Index (DJSI) or Goldman Sachs' SUSTAIN outperformed industry averages by about 15 percent in 16 of 18 categories from May to November 2008.[61]

The financial implications of environmental performance include a lower cost of capital. Sharfman notes Feldman's finding of positive effect of improved environmental risk management on beta and stock price and Garber's outcome of robust positive relationship between environmental liabilities (clearly driven by Superfund fear and notoriety) and costs of equity for large chemical firms. Sharfman found that "firms that develop a strategy to improve their risk management through improved environmental performance reduce their weighted average cost of capital."[62] Although higher levels of environmental risk management permit more debt, they also increase the cost of debt, even after a partial offset by a tax shield. First, lenders see these investments in environmental risk management as inefficient. Second, higher environmental risk management levels are positively related to leverage, which also increases cost of debt.

Bauer and Hann found environmental performance to have similar financial implications on credit risk.[63] They analyzed the profiles of around 600 U.S. public corporations from 1996 to 2006 to test whether environmental concerns are associated with lower credit ratings and higher cost of debt financing. Firms (borrowers) with poor or inadequate management of environmental risks had a higher default risk for bondholders and hence impaired investments, and if faced with significant liability risk, they filed for strategic bankruptcy to avoid the liabilities and in the process also subordinated the claims of their debt holders. This situation was exacerbated by the far-reaching implications of CERCLA, which is retroactive, and has joint and several and strict liabilities for firms and banks. Finally, they observed that the relevance of environmental management issues has increased for bond investors over the recent decade. Their analysis provides comprehensive evidence that corporate environmental management is a cross-sectional credit risk valuation determinant.

3.3.2.2 Environmental, Social, and Governance Performance[64]

A research report by Deloitte shows that long-term business performance not only maximizes financial metrics, but also leverages environmental, social, and governance (ESG) factors.[65] This report supplements financial metrics in understanding how a business is creating value and can throw some light on the long-term sustainability of a business. For instance, investors are increasingly using ASSET4, a Thomson Reuters business information and analysis tool that provides ESG data on over 3,800 companies, and Bloomberg terminals that offer ESG metrics for a more complete picture for the investors. ESG metrics are used by socially responsible investors to sort undesirable companies from preferred companies with strong metrics, and to select companies they want to influence through stock purchase incentives and guide efforts of those companies to direct capital toward underserved areas (e.g., affordable housing). Mainstream investors could use ESG metrics to evaluate company risk elements, incorporate ESG data into traditional financial analysis to get a more complete long-term picture, identify valuable intangibles that may attract investors, and raise share prices.

Ernst & Young suggest that integrated reporting of financial and ESG factors provides a more comprehensive long-term reflection of the value of the firm.[66] It should connect sustainable business practices, tangible and intangible material assets, capital risks and opportunities, and short- and long-term value creation objectives.

3.3.2.3 Sustainability Metrics and CSR Reports

Although not mandatory in most countries, CSR reporting has continued to grow; just over the past three years, of the largest 250 global companies, CSR reporting has grown from 80 to 95 percent, with reputation and ethical considerations topping the list of drivers.[67] As one may expect, companies with higher performance on sustainability metrics are the first ones to report. These could also be the result of a firm's attempt at reducing the cost of capital (see Section 3.3.2.1), and attracting socially responsible, and in particular institutional, investors. Oster observes, "For economists, CSR reports are signals. This is one reason that Apple's opposition to a shareholder petition to require a CSR report in 2009 dismayed some of the firm's supporters."[68] A recent study shows that based on the market value of stocks traded, about a tenth of the publicly listed companies in the United States generate CSR reports. "We find that firms with high cost of equity capital tend to release corporate social responsibility reports.... Further, reporting firms with superior social responsibility performance attract dedicated institutional investors and analyst coverage."[69]

An Ernst & Young survey[70] of 272 sustainability executives in 24 industry sectors with one billion or more revenue showed a lack of internal systems to monitor sustainability impacts and mitigate inaction risks. They note emergence of some key trends and action steps that are presented here:

1. Build same transparency and rigor as financial reporting into sustainability reporting.
2. Engage CFOs in measures to monitor and report on environment and sustainability.
3. Engage employees as key stakeholders to embed sustainability into the corporate culture.
4. Recognize that greenhouse gas disclosure, with growing interest in water, has value outside of the regulatory arena for both internal and external stakeholders.
5. Integrate plans to manage risks in access to key resources in sustainability reporting.
6. Understand the value of sustainability reporting to ranking and rating organizations, particularly those of interest to investors.

These bring into focus two major issues relevant to environmental ethics. First is the need to define consistent format and appropriate and relevant content and an integrated measure. While a plethora of formats and industry sector protocols have been under various stages of development, slowly but surely dominant leading protocols are emerging. Ramanan[71] classifies these as across-the-sector, industry-specific, and national-scale metrics. This section is not an effort at exhaustive coverage of these metrics, and the purpose is to simply highlight some of these.

Launched in 1999, the Dow Jones Sustainability Index (DJSI)[72] tracks financial performance of sustainability-driven companies and provides an integrated economic, social, and environmental assessment to serve as a benchmark for sustainability portfolio asset managers. SAM[73] provides the research backbone for DJSI. Issued in 2000, Global Reporting Initiative (GRI)[74] is a long-term multistakeholder, multinational process to develop and disseminate global guidelines. Into its fourth generation, it is a voluntary guideline for reporting on economic, environmental, and social dimensions of products, activities, and services. DJSI and GRI are two of the dominant elements of the across-the-board sustainability metrics and reporting standards. In recent years, Bloomberg[75] and Thomson Reuters[76] have started providing environmental, social, and most importantly, governance (includes ethics) metrics on their terminals for investors for assessment of nonfinancial factors. This is a very significant development and is a critical play for the environmental professional.

Launched in 2005, the oil and gas sector has developed and follow the oil and gas industry guidance on sustainability reporting,[77] creating a common framework. Similarly, in 2002 the UK Institution of Chemical Engineers launched the sustainability metrics for process industry[78] that can be used within an operating unit. These engineer-developed metrics have absolute, normalized, and qualitative measures. These are representative of industry sector-specific sustainability reporting metrics covering two of the largest and highest sustainability impact categories.

Finally, World Economic Forum[79] has developed national scale sustainability metrics for nations to be compared on their environmental performance. It is a depository of environmental data.

In addition to the development of standards for the sustainability metrics reported, efforts are being made by the international community to ensure consistency and reliability of the reported information. These are presented in Section 3.3.2.4. Many leading companies are moving toward integrated reporting, for instance, CLP Holdings.[80] Integrated reporting has also grown almost fivefold in the past five years globally.

Second is the requirement to protect society at large from the avalanche of unethical green washing. Here in this section, we will further highlight the risks of green washing and how to mitigate them. Demas and Burbano[81] define green washing as the intersection of two behaviors—poor environmental performance and positive communication about environmental performance—and go on to expand on it as the act of misleading consumers regarding the environmental practices of the company or the environmental performance of the product or service. At the product level, green-washing sins may involve claiming benefits with no evidence, or involve a trade-off with some other negative impact that is not disclosed. They identify several groups of drivers. The first group is market based, with pressure coming from competitors or from consumer or investor demand. Organizational culture, incentive structure, or internal pressures form the second group. Individual psychological drivers include individual bias or narrow framework. Finally, the nonmarket group comprises a lax regulatory environment and lack of monitoring by activists and media. They make some valuable recommendations that could be critical from an environmental ethics perspective. They suggest transparency in environmental performance disclosure, whether voluntary or mandatory; creating awareness through sharing information about incidents of green washing and by reducing regulatory uncertainty; and most significantly providing ethical leadership and training for employees.

3.3.2.4 International Integrated Reporting Community (IIRC)

The International Integrated Reporting Community[82] has stated the following guiding principles for its proposed enhancement of the integrated financial and sustainability reporting standards:

Guiding Principle 1 (Strategic focus and future orientation): An integrated report should provide insight into the organization's strategy and how that relates to its ability to create value in the short, medium, and long term, and to its use of and effects on the capitals.

Guiding Principle 2 (Connectivity of information): An integrated report should show, as a comprehensive value creation story, the combination, inter-relatedness, and dependencies between the components that are material to the organization's ability to create value over time.

Guiding Principle 3 (Stakeholder responsiveness): An integrated report should provide insight into the quality of the organization's relationships with its key stakeholders and how and to what extent the organization understands, takes into account, and responds to their legitimate needs, interests, and expectations.

Guiding Principle 4 (Materiality and conciseness): An integrated report should provide concise information that is material to assessing the organization's ability to create value in the short, medium, and long term.

Guiding Principle 5 (Reliability): The information in an integrated report should be reliable. The reliability of information is affected by its completeness, neutrality, and freedom from error.

Guiding Principle 6 (Comparability and consistency): The information in an integrated report should be presented in a way that enables comparison with other organizations to the extent it is material to the reporter's own value creation story, and on a basis that is consistent over time.

3.3.3 Building an Ethical Culture

Top management, including the board of directors, must recognize the link between ethics, long-term profits, and clear policies on ethics that promote leaders based on performance and ethical conduct and create a safe environment for whistleblowers to report irregularities. Leadership's role in building an ethical culture starts with value statements that connote integrity, doing the right thing, and genuine concern for stakeholders, and a commitment to a code of ethics that serves as a basis for an organization's conscience. At the operating level, senior management must have procedures in place to employ and partner only with those who meet ethical standards, and to ensure middle management fosters compliance with ethical standards and helps convince employees to spot and report ethical errors. As in most culture-building endeavors, success will depend on consistent and effective communication through both formal and informal channels—word of mouth indoctrination to instill a power of tradition, annual training through participatory workshops, and global communication through annual reports, websites, and orientation.

Organizational commitment and visible leadership behavior are basic to the institutionalization of an ethical culture. To build an ethical culture, it is mandatory that leadership take personal responsibility backed by a visible, explicit, unequivocal, and irreversible commitment. Leadership must promote and model the use of these values in making decisions and actions and eliminate autonomous fiefdoms within management. Ethics must be formally and explicitly built into daily business life. Short-term, organization-wide communication in the form of public statements is critical. Ethics must be seen as expected and important, on par with legal, financial, and other regulatory compliance. Longer term, displays of ethical values must be encouraged and the environment should promote appropriate intrepid whistleblowing, in order to build and internalize ethics as part of the organization's culture. It is also critical that the use of ethical factors and alternatives in decisions be explicitly and publicly explained at the operational level. It must be understood and accepted as not just lip service.

3.3.3.1 Corporate Governance and Global Stewardship

The definition of corporate governance adopted by the Organization for Economic Cooperation and Development (OECD) is broad and states that corporate governance "is the system by which business corporations are directed and controlled."[83] The OECD's Principles of Corporate Governance, first published in 1999 and updated in 2004 after the spate of corporate scandals, contains strong recognition of the importance of stakeholders and the need for timely, accurate, and transparent disclosure of all relevant and material information, and in particular includes ethics and the environment, as stated below:

> In addition to their commercial objectives, companies are encouraged to disclose policies relating to business ethics, the environment and other public policy commitments. Such information may be important for investors and other users of information to better evaluate the relationship between companies and the communities in which they operate

> and the steps that companies have taken to implement their objectives.... In fulfilling its control oversight responsibilities it is important for the board to encourage the reporting of unethical/unlawful behavior without fear of retribution. The existence of a company code of ethics should aid this process which should be underpinned by legal protection for the individuals concerned.[84]

In particular, it includes provision for ways to report unethical behavior and legal protection for the whistleblower. High-integrity standards restore public and investor trust and help engage stakeholders, especially communities, in directing business performance as regards fulfilling the new social contract.[85] The use of the Dow Jones Sustainability and FTSE4Good Indexes for choosing corporations to invest in by the growing portfolio of socially responsible investors, and the use of the Equator Principles to assess social and environmental risk for project financing by many major financial institutions, including the World Bank, confirm that socially responsible practices can improve access to financial markets and reduce capital costs. Good corporate governance is instrumental in building an ethics culture.

3.3.3.2 Emotional Power of Leadership Commitment

> An adequate corporate strategy must include noneconomic goals.... An economic strategy is humanized and made attainable in a living organization by deciding on the character the company is to have, the values it espouses, and its relationships to its customers, employees, communities, and shareholders. The personal values and ethical aspirations of the company leaders, though probably not specifically stated, are implicit in all strategic decisions.... Although codes of ethics, ethical policy for specific vulnerabilities, and disciplined enforcement are important, they do not contain in themselves the final emotional power of commitment. Commitment to quality objectives—among them compliance with law and high ethical standards—is an organizational achievement. It is inspired by pride more than the profit that rightful pride produces. Once the scope of strategic decision is thus enlarged, its ethical component is no longer at odds with a decision right for many reasons.[86]

This statement very effectively captures the significance of personal values and ethical standards, as well as the emotional power of commitment of company leaders in building an ethics culture in the organization. For instance, 3M pioneered the concept of pollution prevention with the creation of the Pollution Prevention Pays (3P) program in 1975. This was made possible, based on private conversations the author had with senior executives in 3M, only because of the vision and commitment of the company leader. The driver behind the successful launch of Walmart's now well-known greening of the supply chain was the emotional power of commitment of then-CEO Lee Scott's vision. Tata, the leading industrial powerhouse of India and the fifth largest steel company in the world, has, for over a century, exhibited impeccable ethical culture, maintaining a high ethical standard throughout its operations in India and globally, driven by the Tata family's commitment.

3.3.3.3 Legal versus Ethical Compliance

We want to know why we are making any decision the way we are, especially ethical ones that bring in our own value systems. The "because Mommy said so" approach works only up to a very tender age. Hence, simply complying with legal dictates because it is our duty is not the go-to solution for most decision makers. It is seen as an obligation imposed by an external source. Ethical decision making, especially after one is well trained in the process and loses the fear of the unknown, provides a sense of freedom, and also promotes a desire to accomplish. This is also consistent with the concept of organizational achievement and sense of pride discussed in the previous section. The legal approach uses rules, monitoring, and deterrents to enforce; the ethical method permeates into every level of decision, all the way from upper management to where the rubber meets the road. Arjoon observes that the legal approach promotes a rule-based "stick" approach and fails to inspire; on the other hand, the ethical approach is more akin to the spirit of the law and its principle-based "carrot" approach is more effective.[87] More significantly, the legalistic approach is dictated by a checklist-based decision-making process. These aspects bring to mind a frequent debate encountered by the coauthor in his corporate experience. Often, the less experienced managers would question the value of bringing costly internal experts to evaluate complex environmental and safety issues. They would ask, "Why can't we just get a checklist I could go through?" One would clearly lose sight of the right thing to do. Once again, this highlights the value of participatory ethics workshops to train new managers and keep experienced ones current in their ethical decision-making processes.

3.3.3.4 Warren Buffett and Salomon Brothers

Warren Buffett's turnaround of Salomon Brothers provides a classic example of the value of ethical compliance beyond just legal compliance and how to restore organizational reputation after catastrophic integrity failure.

The steps to restore reputation, especially after a catastrophic failure of integrity, are succinctly captured by Fombrun and noted by Sims:

1. Take public responsibility
2. Convey concern to all stakeholders
3. Full cooperation with authority
4. Remove negligent incumbents
5. Appoint credible leaders that represent all interests
6. Dismiss tied suppliers/agents
7. Hire independent investigators

8. Reorganize for greater control
9. Establish strict procedures
10. Eliminate infraction practices
11. Revise practices and pay systems
12. Monitor compliance[88]

Sims describes how the (lack of) ethics culture at Salomon Brothers was turned around.[89] Warren Buffett, when he took over as CEO of Salomon Brothers in 1991, crafted a new corporate culture in order to resurrect the company. As a first step, he reached out to his employees with clear messages. For instance, he acknowledged to the entire workforce that "in some way, we had lost our way.... A bravado was attached to the taking of risk and the making of money. As a result we were inattentive to shareholders and external constituencies."[90] Next, he announced across the organization that "you are each expected to report, instantaneously and directly to me any legal violation or moral failure on behalf of any employee of Salomon.... You are to make reporting directly to me your first priority."[91] He also made it crystal clear that not only should one not be afraid of blowing the whistle, but it is in fact a mandate that one do so.

Buffett also went public with a compliance note to major print media like *Wall Street Journal, New York Times, Financial Times*, etc. He told them, "[We will] be guided by the test that goes beyond the rules ... employee should ask oneself if one would be willing to see on the front page of the local newspaper, there to be read by his/her spouse, children and friends ... simply want no part of any activities that pass legal tests but that we as citizens, would find offensive."[92]

3.3.3.5 Ethics Monitoring and Enforcement

Ethics monitoring and enforcement is another decisive factor for the continued success of institutionalizing and building an ethics culture in any organization. In addition to the distribution of written codes, policies, and guidelines to all concerned parties, reinforcement through broad communication by executives, and assigning advisors, hotlines, and ombudsmen, it is essential to develop training and implement programs geared to help apply policy to everyday work situations. This is best accomplished through frequent participatory workshops.

Ethical values, like most others, cannot be taught; rather, they must be lived. Employees will do what they see. Values must be simple, easy to articulate, realistic, and applicable; they must apply to internal and external operations. These are first and best communicated in the selection of employees. Ideally, values should be in agreement with the diverse stakeholders, show obsession/passion with fairness and equity, and seek not just collective, but individual responsibility.

3.3.3.6 Ethics Filter in Business Decisions

This book follows Justice Potter Stewart's definition of ethics as the difference between what a person has the right to do and the right thing to do (see Section 1.2.1). As pointed out earlier, this definition is consistent with Paine's[93] elegant distinction, and that ethical decisions go beyond the legal tenet of limits to stay within. Arjoon observes that ethics is more akin to the spirit of the law and inspires excellence.

In the earlier section we discuss the evolution of the framework for making ethical decisions. The simplest ethics filter to apply in everyday life may be the basic yet robust four-way test used since 1932 by the Rotary Club,[94] a charitable organization. Is it true and fair, and will build goodwill and benefit all. Nearly half a century later, Nash[95] set out a dozen pragmatic queries for business executives (see Section 3.2.4.2). Another quarter century later Paine[96] transformed those into eight business-specific ethics principles (see Section 3.2.4.4). Sucher[97] suggests evaluating each action from multiple ethical perspectives.

British Standards Institution[98] presents a framework and toolkit for creating a business case for sustainability. The five-step process from a company perspective is as follows:

1. Understand your significant impacts on the environment, society, and the economy and identify opportunities and risks they represent—both real and perceived.
2. Identify key stakeholder issues through consultation—both real and perceived—to improve ability to maximize opportunities and minimize risks.
3. Make it relevant; that is, link the opportunities and risks to the company's core business and map the issues and impacts to the company business plan and strategic objectives.
4. Back it up with data and examples to support each opportunity and risk; in particular, include available or estimated financial costs and benefits.
5. Keep it dynamic and updated—ensure it changes with company priorities and keep it relevant for the audience.

Ramanan[99] suggests applying that approach to building ethics filters as well in the corporate decision process by applying the eight ethics principles that Paine suggests. See Section 3.2.4.4 for an expansion of these business-specific ethical principles in the context of environmental ethics.

Finally, Ethics Resource Center[100] presents the Plus method to test all corporate decisions for meeting the ethics threshold. The process uses the mnemonic PLUS to reflect policies and procedures of the corporation, laws and regulations that apply, universal ethical principles that the corporation has adopted, and the self-defined values of the individual making the decision. It suggests surfacing ethical issues in defining any problem, assessing the ethical impacts in evaluating identified alternatives, and resurfacing any remnant ethical issue in the selected option.

3.3.4.7 Taking Ethics Responsibility Global

Kanter says that companies and leaders will be judged by the ultimate effect of their action on societal well-being. Under the new social contract companies will have to thrive under the influence of multiple stakeholders and businesses.[101] Paine recommends a broad, dynamic, and affirmative profile of corporate conduct, calling for a flexible shift from "zero tolerance" to noncompliance.[102] This recent paper suggests a new way of looking at corporate ethics. Broad implies standards for imperfect duties that incorporate significant degrees of freedom. It is difficult to state these in legal auditable terms. Dynamic captures the shift in behavior over time, changing with the macroeconomic status and industry conditions. Affirmative refers to ethics standards as goals to strive for rather than strict lapses to avoid.

Endnotes

1. Ram Ramanan and W. Ashton, "Green MBA and Integrating Sustainability in Business Education," *Environmental Manager*, September 2012, pp. 13–15; also accessed December 2012, http://stuart.iit.edu/about/faculty/pdf/green_mba.pdf.
2. Various management consultant reports on progress of sustainability:
    a. KPMG, *Building Business Value in a Changing World*, accessed December 2012, http://www.kpmg.com/Global/en/IssuesAndInsights/Articles Publications/Documents/building-business-value.pdf.
    b. KPMG, *Corporate Sustainability: A Progress Report*, accessed December 2012, http://www.kpmg.com/Global/en/IssuesAndInsights/Articles Publications/Documents/corporate-sustainability-v2.pdf.
    c. MIT Sloan Management Review and the Boston Consulting Group, "Sustainability Nears a Tipping Point," Research Report (Winter 2012).
3. Business for Social Responsibility, "Bloomberg Launches ESG Metrics," last modified November 17, 2009, https://www.bsr.org/our-insights/bsr-insight-article/bloomberg-launches-esg-data-service.
4. M.E. Porter and Mark R. Kramer, "Creating Shared Value," *Harvard Business Review*, January–February 2011, pp. 62–77.
5. Ramanan, "Green MBA."
6. Ibid.
7. Ram Ramanan, "Need for Green MBA and Environmental Economics Education—Globally" (AWMA Annual Conference, Portland, OR, June 2008).
8. Ramanan, "Green MBA."
9. M.E. Porter and M.R. Kramer, "Strategy and Society," *Harvard Business Review*, December 2006, pp. 78–88.

10. Effective Crisis Management, "The Tylenol Crisis," accessed December 2012, http://iml.jou.ufl.edu/projects/fall02/susi/tylenol.htm.
11. Aneel Karnani, "Case against CSR," *Wall Street Journal*, August 23, 2010.
12. UN Brundtland Commission, Report of the World Commission on Environment and Development: *Our Common Future*, UN Brundtland Commission (United Nations, 1987).
13. Ramanan, "Green MBA."
14. Porter, "Strategy and Society."
15. Karnani, "Case against CSR"; coauthor Ramanan private conversation with son Shivraj Ramanan (who was a student of Karnani at the Ross School of Business at the University of Michigan).
16. Wayne Visser, "Ages and Stages of CSR," in *The Age of Responsibility*, 1st ed. (Hoboken, NJ: John Wiley & Sons, 2011), p. 21.
17. Wikipedia, s.v. "Greed Is Good," accessed December 2012, http://en.wikipedia.org/wiki/Gordon_Gekko#.22Greed_is_Good.22_quotation.
18. Visser, "Ages and Stages," p. 37.
19. Bill Clinton in his Foreword in Matthew Bishop and Michael Green, *Philanthrocapitalism—How Giving Can Save the World* (New York: Bloomsbury Press, 2008).
20. Daniel Franklin, ed., "The Economist's World in 2009 Yearbook," in Wayne Visser, "Ages and Stages of CSR," in *The Age of Responsibility* (Hoboken, NJ: John Wiley & Sons, 2011), p. 94.
21. Archie Carroll, "The Pyramid of Corporate Social Responsibility," *Business Horizons* 42 (1991): 39–48.
22. Visser, "Ages and Stages," p. 131.
23. C.K. Prahlad and Stuart L. Hart, *The Fortune at the Bottom of the Pyramid* (Upper Saddle River, NJ: FT Press, 1998); G. Hamel and C.K. Prahlad, *Competing for the Future* (Boston: Harvard Business School Press, 1994).
24. C.K. Prahlad and Stuart L. Hart, "The Fortune at the Bottom of the Pyramid," Strategy and Business 26 (2002), accessed December 2012, http://www.cs.berkeley.edu/~brewer/ict4b/Fortune-BoP.pdf.
25. Aneel Karnani, "Eradicating Poverty through Enterprise" (presented at the University of Michigan, November 2007), accessed December 2012, http://www.un.org/esa/coordination/Eradicating%20Poverty%20through%20Enterprise.Karnani.ppt; Aneel Karnani, *Fighting Poverty Together: Rethinking Strategies for Business, Governments, and Civil Society to Reduce Poverty* (London: Palgrave Macmillan, 2011), p. 44.
26. Amartya Sen, Nobel laureate economist quoted by Aneel Karnani at the United Nations General Meeting in New York, 2007, accessed December 2012, http://www.un.org/esa/coordination/Eradicating%20Poverty%20through%20Enterprise.Karnani.ppt.
27. Karnani, *Fighting Poverty Together*, p. 232.
28. M.E. Porter and Mark R. Kramer, "Creating Shared Value"; Ram Ramanan, "How to Build Sustainability Issues in Corporate Decision Making" (lecture, Net Impact Student Chapter Seminar, IIT Start School of Business, Chicago, March 25, 2011).
29. Ram Ramanan, "How to Build Sustainability Issues in Corporate Decision Making" (Net Impact Student Chapter Seminar, IIT Stuart School of Business, Chicago, March 25, 2011).

30. Meg Voorhes et al., "Executive Summary—Fig. B: Growth of SRI $2.7 Trillion in 2007 to $3.0 Trillion in 2010," in *2010 Report on Socially Responsible Investing Trends in the United States*, Social Investment Forum Foundation, accessed December 2012, http://ussif.org/resources/research/documents/2010TrendsES.pdf.
31. Sims, *Corporate Social Responsibility*, p. 99.
32. Ronald R. Sims and Johannes Brinkmann, "Leaders as Moral Role Models: The Case of John Gutfreund at Salomon Brothers," *Journal of Business Ethics* 35, no. 4 (2002): 327–339; R. Reidenbach and D.P. Robin, "A Conceptual Model of Corporate Moral Development," *Journal of Business Ethics* 10, no. 4 (1991): 273–284.
33. Carroll, "Pyramid of Corporate," pp. 39–48.
34. Porter, "Strategy and Society."
35. Porter, "Creating Shared Value"; Visser, "Stages of CSR."
36. Wikipedia, s.v. "Groupthink," accessed December 2012, http://en.wikipedia.org/wiki/Groupthink.
37. J. Haidt and S. Kesebir, "Morality," in *Handbook of Social Psychology*, 5th ed., ed. S. Fiske and D. Gilbert (Hoboken, NJ: John Wiley & Sons, 2010), pp. 797–832.
38. Haidt, *The Righteous Mind*, p. 187.
39. I.L. Janis, *Victims of Groupthink* (Boston: Houghton-Mifflin, 1972), p. 197; Sims, "Corporate Social Responsibility," p. 117.
40. Herbert J. Taylor, "Rotary International Code of Ethics," Rotary International, accessed December 2012, http://www.rotary.org/en/aboutus/history/rihistory/pages/ridefault.aspx.
41. Dorothy Frede, "Plato's Ethics: An Overview," in *Stanford's Encyclopedia of Philosophy*, ed. Edward N. Zalta, accessed December 2012, http://plato.stanford.edu/entries/plato-ethics/#VirStaSou.
42. Laura L. Nash, "Ethics without the Sermon," *Harvard Business Review* 59 (1981): 79–89.
43. Equator Principles, "Environmental and Social Risk Management for Project Finance," accessed December 2012, http://www.equator-principles.com/; United Nations, "Overview of the UN Global Compact," accessed December 2012, http://www.unglobalcompact.org/aboutthegc/thetenprinciples/index.html.
44. Hal Taback, "Ethics Training: An American Solution for Doing the Right Thing," in *Engineering and Environmental Ethics*, ed. John Wilcox and Louis Theodore (New York: John Wiley, 1998), pp. 267–280.
45. Lynn S. Paine, Rohit Deshpande, Joshua D. Margolis, and Kim E. Bettcher, "Up to Code: Does Your Company's Conduct Meet World-Class Standards," *Harvard Business Review* 83, no. 12 (2005).
46. Sandra J. Sucher, *Teaching the Moral Leader: A Literature-Based Leadership Course: A Guide for Instructors* (London: Routledge, Taylor and Francis, 2012) p. 110; Sandra J. Sucher, *The Moral Leader: Challenges, Tools and Insights* (London: Routledge, 2008).
47. Keith T. Darcy, "The Last Decade" (presented at Business Roundtable, Institute for Corporate Ethics, Charlottesville, VA, November 2009).
48. Robert F. Bruner, "The Economic Climate's Impact on Corporate Culture and Ethics" (presented at Business Roundtable, Institute for Corporate Ethics, Charlottesville, VA, November 2009).
49. Rachel Carson, *Silent Spring* (Boston: Mariner Books, 2002), accessed December 2012, http://www.goodreads.com/work/quotes/880193-silent-spring.

50. Dan Currell, "Weathering the Integrity Recession" (presented at Business Roundtable, Institute for Corporate Ethics, Charlottesville, VA, November 2009).
51. Business Roundtable, Institute for Corporate Ethics, *2009 Compliance and Ethics Forum Summary Report: Leading Thoughts and Practices*, November 2009.
52. John J. Castellani, "The Impact of the Economic Climate on Corporate Culture" (presented at Business Roundtable, Institute for Corporate Ethics, Charlottesville, VA, November 2009).
53. Currell, "Integrity Recession."
54. Sims, *Corporate Social Responsibility*, p. 7.
55. Rosabeth Moss Kanter, "It's Time to Take Full Responsibility," *Harvard Business Review*, October 2010, p. 1.
56. Equator Principles, "History of Equator Principles," *Equator Principles*, accessed December 2012, http://www.equator-principles.com.
57. BankTrack, *Collevecchio Declaration* (Amsterdam: BankTrack, January 2003), accessed December 2012, http://www.banktrack.org/download/collevechio_declaration/030401_collevecchio_declaration_with_signatories.pdf.
58. European Commission Initiative for Mandatory Environmental, Social and Governance Disclosure in the European Union.
59. World Business Council for Sustainable Development (WBCSD), "Business Solutions for a Sustainable World," accessed December 2012, http://www.wbcsd.org/.
60. Strandberg Consulting, *The Business Case for Sustainability*, accessed December 2012, http://www.corostrandberg.com/pdfs/Business_Case_for_Sustainability_21.pdf.
61. SAM Sustainability Investing, "Dow Jones Sustainability Indexes," accessed December 2012, http://www.sustainability-index.com/; Goldman Sachs, SUSTAIN, "GS SUSTAIN," December 2012, http://www.goldmansachs.com/our-thinking/topics/gs-sustain/index.html.
62. Mark Sharfman and Fernando Chitru, "How Does Overall Environmental Risk Management Affect the Cost of Capital," *Strategic Manual Journal* 29, no. 6 (2008): 569–592, doi: 10/1002/smj.678; Stanley J. Feldman, Peter A. Soyka, and Paul G. Ameer, *Does Improving a Firm's Environmental Management System and Environmental Performance Result in a Higher Stock Price? Environmental Group Study* (Fairfax, VA: ICF Kaiser International, 1996), *Journal of Investing* 6, no. 4 (1997): 87–97; S. Garber and J.K. Hammitt, "Risk Premiums for Environmental Liabilities: Superfund and the Cost of Capital," *Journal of Environment and Economic Management* 36 (1998): 267–294.
63. Rob Bauer and Daniel Hann, *Corporate Environmental Management and Credit Risk*, Maastrict University, European Centre for Corporate Engagement, last modified June 2010, http://www.responsible-investor.com/images/uploads/Bauer__Hann_%282010%29.pdf.
64. Thomas Reuters, *Corporate Responsibility Report*, accessed December 2012, http://thomsonreuters.com/about/corporate_responsibility/ESG_performance/; Peter A. Soyka and Mark E. Bateman, *Finding Common Ground on the Metrics That Matter*, IRRC Institute, last modified February 2012, http://www.irrcinstitute.org/pdf/IRRC-Metrics-that-Matter-Report_Feb-2012.pdf.
65. Bart King, "Deloitte: ESG Metrics Crucial for Creating Business Value," Sustainable Brands, last modified April 3, 2011, http://www.sustainablebrands.com/news_and_views/articles/deloitte-reports-esg-metrics.

66. Brendan LeBlanc, Benjamin Miller, and Jeremy Osborn, *Driving Value by Combining Financial and Non-Financial Information into a Single, Investor Grade Document*, Ernst & Young, accessed December 2012, http://www.ey.com/Publication/vwLUAssets/Integrated_reporting:_driving_value/$FILE/Integrated_reporting-driving_value.pdf.
67. KPMG, "State of Corporate Responsibility Reporting," accessed December 2012, http://www.kpmg.com/PT/pt/IssuesAndInsights/Documents/corporate-responsibility2011.pdf.
68. Sharon Oster, "What Do We Make of CSR Reporting?" accessed December 2012, http://www.forbes.com/sites/csr/2010/05/11/what-do-we-make-of-csr-reporting/.
69. Dan S. Dhaliwal, Oliver Zhen Li, Albert Tsang, and Yong George Yang, Voluntary Non-Financial Disclosure and the Cost of Equity Capital: The Case of Corporate Social Responsibility Reporting, February 15, 2009, http://ssrn.com/abstract=1343453 or http://dx.doi.org/10.2139/ssrn.1343453.
70. Ernst & Young, "Six Growing Trends in Corporate Sustainability," accessed December 2012, http://www.ey.com/US/en/Services/Specialty-Services/Climate-Change-and-Sustainability-Services/Six-growing-trends-in-corporate-sustainability_overview; http://www.ey.com/US/en/Services/Specialty-Services/Climate-Change-and-Sustainability-Services/Six-growing-trends-in-corporate-sustainability_Trend-7.
71. R. Ramanan, "Environmental Performance Reporting—State of the Art" (Proceedings of International Interdisciplinary Conference on Sustainable Technologies for Environmental Protection, ICSTEP2006, Coimbatore, India).
72. Dow Jones, "Dow Jones Sustainability Indexes," accessed December 2012, http://www.sustainability-index.com/.
73. http://www.sustainability-index.com/images/sam-csa-methodology-en_tcm1071-338252.pdf.
74. Global Reporting Initiative, accessed December 2012, https://www.globalreporting.org (coauthor Dr. Ramanan is a member of the GRI's G4 Academic Research Group).
75. Bloomberg, "Sustainability," accessed December 2012, http://www.bloomberg.com/bsustainable/.
76. Thomson Reuters (ASSET4), "ESG/CSR Content Overview," accessed December 2012, http://thomsonreuters.com/products_services/financial/content_news/content_overview/content_az/content_esg/.
77. IPIECA (the global oil and gas industry association for environmental and social issues), API (the American Petroleum Institute), and OGP (the International Association of Oil and Gas Producers), Oil and Gas Industry Guidance on Voluntary Sustainability Reporting, 2nd ed., 2010, accessed December 2012, http://www.api.org/environment-health-and-safety/~/media/files/ehs/environmental_performance/voluntary_sustainability_reporting_guidance_2010.ashx.
78. Institution of Chemical Engineers, UK, "Sustainability Metrics for the Process Industry," accessed December 2012, http://www.icheme.org/communities/special-interest-groups/sustainability/~/media/Documents/Subject%20Groups/Sustainability/Newsletters/Sustainability%20Metrics.ashx.
79. World Economic Forum, "Sustainability Data and Trends," accessed December 2012, http://sedac.ciesin.columbia.edu/theme/sustainability.

80. Private conversations with Jeanne Ng, Director of Environmental Affairs, CLP Holdings; KPMG, "Corporate Sustainability: A Progress Report," April 2011, accessed December 2012, http://www.kpmg.com/Global/en/IssuesAndInsights/ArticlesPublications/Documents/corporate-sustainability-v2.pdf.
81. M.A. Demas and V.C. Burbano, "The Drivers of Green-Washing," California Management Review 54, no. I (2011), CMR494.
82. International Integrated Reporting Community, accessed April 2013, http://www.theiirc.org/wp-content/uploads/2012/11/23.11.12.Prototype-Final.pdf.
83. *OECD Principles of Corporate Governance* (Paris: OECD Publications, 2004), accessed December 2012, http://www.oecd.org/daf/corporateaffairs/corporategovernanceprinciples/31557724.pdf.
84. Bart, "ESG Metrics."
85. Global Corporate Governance Forum and UN Global Compact, *Corporate Governance: The Foundation for Corporate Citizenship and Sustainable Business*, 2009, http://www.unglobalcompact.org/docs/issues_doc/Corporate_Governance/Corporate_Governance_IFC_UNGC.pdf,
86. Kenneth Andrews, "Ethics in Practice: Managing the Moral Corporation," *Harvard Business Review*, September-October 1989, p. 6. With permission.
87. Surendra Arjoon, "Corporate Governance: An Ethical Perspective," *Journal of Business Ethics* 61, no. 4 (2005): 343–352.
88. Charles J. Fombrun, *Reputation: Realizing Value from the Corporate Image* (Boston: Harvard Business School Press, 1996), presented in Sims, *Corporate Social Responsibility*, p. 234.
89. Sims, *Corporate Social Responsibility*, p. 211.
90. Ibid., p. 221.
91. Ibid., p. 220.
92. Ibid., p. 223.
93. Lynn Sharp Paine, *Venturing beyond Compliance: The Evolving Role of Ethics in Business* (New York: The Conference Board, Inc., 1996), 13–16.
94. Herbert J. Taylor, "Rotary International Code of Ethics."
95. Laura L. Nash, "Ethics without the Sermon."
96. Lynn S. Paine, "Up to Code."
97. Sandra J. Sucher, *Teaching the Moral Leader*.
98. British Standards Institution, "The Sigma Guidelines Toolkit," accessed December 2012, http://projectsigma.co.uk/toolkit/sigmabusinesscase.pdf.
99. Ram Ramanan, "How to Build Sustainability Issues in Corporate Decision Making" (Net Impact Student Chapter Seminar, IIT Start School of Business, Chicago, March 25, 2011); Ram Ramanan, "How to Build Ethics Filter in Corporate Decision Making" (104th Air and Waste Management Association Annual Conference and Exposition, Orlando, June 23, 2011).
100. Ethics Resource Center, "Ethics Filter," accessed December 2012, http://www.ethics.org/resource/ethics-filters.
101. Kanter, "Full Responsibility."
102. Lynn S. Paine, Rohit Deshpande, and Joshua Margolis, "A Global Leader's Guide to Managing Business Conduct," *Harvard Business Review* 89, no. 9 (2011) (online edition).

4

Environmental Ethics Training for Leaders and Professionals

Norman R. Augustine, retired chairman and CEO of Lockheed Martin, gave one of the most valuable explanations of how unethical behavior unfolds in his Pace Award acceptance speech. The Pace Award is presented annually by the Ethics Resource Center in Washington, D.C., to honor individuals and groups from the corporate world who display excellence in the field of ethics.[1] Augustine and his fellow colleagues at ERC bring us hope in the wake of the highly publicized corporate scandals of recent years.

When it comes to corporate action, Augustine says, "There can be nothing more important in life than ethics and the closely related matter of one's reputation. These are even more important than one's health." Why then do seemingly decent people make such unethical decisions? In most cases, wrongdoing is not the result of a single event or wrong turn, but rather a series of wrong turns. Augustine compares ethical collapses to that of a boa constrictor devouring its prey. Most believe that a boa constrictor simply wraps itself around its prey and crushes it to death in a single motion. On the contrary, Augustine points out that the boa waits once it has wrapped up its victim. Each time the prey exhales, the boa tightens its grip until, eventually, the prey can no longer breathe. This explains why it is so important that supervisors be alerted to minor ethical breeches in the workplace, such as cheating on expense accounts, overcharging client hours, or using company property for personal use. Augustine says, "Each event moves the person closer to a major ethical transgression."[2]

4.1 Building an Environmental Ethics Training Program

Ethics means doing the right thing (as opposed to what you have the right to do and definitely not doing anything that you do not have a right to do). But doing the right thing is not always obvious or easy. In fact, ethical decisions are often difficult and may involve a certain amount of self-sacrifice. Also, while it sounds simple, the "right thing" is not always obvious. If a situation arises that we have not faced in the past, we need to be prepared to react appropriately. Many believe that our instincts will lead us to do the right thing. Although unfortunate, even "good managers often make unethical decisions—and don't even know it."[3] Ethical decision making, by definition, is difficult. In fact, if the decision is not difficult, it is probably not an ethical dilemma. Ethical decisions involve an element of self-sacrifice. The sacrifice might involve losing a client or a job. The sacrifice can involve denying your family because your job is at stake if you resist taking an action that would threaten public health or violate the law. Issues come up unexpectedly, and often a person with the best of intentions makes the wrong decision.

It often takes training to even recognize an ethical dilemma, so doing the right thing requires training. In many respects, learning to behave ethically is similar to learning any behavior—from language skills to athletics. Just as no person can jump out of the grandstand and compete with an Olympic runner, no matter how physically fit that person is, performing ethically requires continuous training. Ethics training, like any formal training, consists of theory and practice, like lectures and lab work, and involves learning what is right and then practicing it on a regular basis to achieve perfection. Overcoming the human instinct toward selfishness, which can have such a strong influence on ethical behavior, requires just as much practice and discipline as overcoming the body's physical limitations in achieving athletic success. One can read and watch videos on hitting a baseball or golf ball, playing tennis, swimming, or performing gymnastics, but perfection comes only with repeated practice. Baseball players play nearly every day, but still practice batting before each game, and professional golfers often hit a bucket of balls before each round in a tournament. It is not sufficient to have a one-time lecture of do's and don'ts regarding professional behavior and expect the professional to perform correctly and do the right thing in tough situations that involve a risk to his or her own comfort and security as well as that of his or her loved ones. It is also insufficient for a firm to produce a code of ethics and have employees sign a certification indicating that they have read and understand that code.

The primary responsibility of an environmental professional or manager is to protect public health and safety, but he or she also has a responsibility to his or her family and career, employer/client, and the environment. Meeting those responsibilities will challenge environmental professionals to draw on their system of ethical values and priorities. Doing the right thing for an environmental professional can be especially challenging. Because ethical dilemmas occur infrequently, environmental professionals are often unprepared to handle them when they do arise. The ethical aspects of a situation may even go unrecognized due to an eroding or faulty value system. The environmental professional must be educated and, more importantly, trained through frequent participatory workshops with real-world scenarios to be able to make the right choice when faced with environmental dilemmas.

4.1.1 Evolution of Environmental Ethics Training

Ethics training for environmental professionals evolved over a 20-year period starting in the late 1980s, and originated in conjunction with the Air and Waste Management Association (A&WMA).[4] At that time environmental professionals were heavily involved with the requirement to audit industrial operations. Consultants were hired by corporate headquarters to evaluate existing facilities and potential plant sites around the United States. The coauthor of this book, while leading an audit/consulting team, became aware that some staff members were not always reporting deficiencies when the plant manager requested they not be reported if the deficiency was immediately corrected. This was the genesis of ethics training workshops, started by the coauthor, to ensure that those on his audit team understood their responsibility and learned how to handle such requests with the smallest impact on various stakeholders possible (i.e., the client, the consulting firm, the individuals involved with the audits). This training concept was originally raised at an annual meeting of the A&WMA held in Anaheim, California, in June 1988, and an A&WMA ethics committee was formed that has functioned for over 20 years. A column was published by the coauthor of this book from 2002 through 2008 in *EM* magazine entitled "Ethics Corner," in which specific ethical dilemmas and ways to address them were presented with associated rationales.[5] Many of the dilemmas used in that column are presented later in this book.

4.1.2 Evolution of Code of Environmental Ethics

In the late 1980s the coauthor presented workshops on ethics to national, regional, and local gatherings of environmental and other technical professionals in the United States and Canada. The feedback from those workshops led to the development of a code of ethics for A&WMA that formed the basis for similar ethics codes for the American Academy of Environmental Engineers and the Institute of Professional Environmental Practice.[6] Around the same time, a few leading environmental consultants, such as AECOM (formerly ERT), for which the other coauthor was an environmental practice leader, had a code of ethics that was mandatory for all employees.[7] As part of its code of ethics, A&WMA developed a set of priorities that environmental professionals can use to solve ethical dilemmas and arrive at a decision for doing the right thing. The foremost responsibility of an environmental professional is to protect public health and the environment. Next in importance is family and career, followed by employer and client. A&WMA's code of ethics was formulated to state the priorities and responsibilities of an environmental professional in detail, and it provides a sound basis for dealing with a manager or client who might risk violating the public health priority. A&WMA's code of ethics, developed by a committee comprised of environmental professionals, is a good example (see the appendix).

4.1.3 Training to Make Code of (Environmental) Ethics Operative

Today many companies have a comprehensive code of ethics that includes specific ethical breeches to be avoided. For a large company, the code is broad and relevant to practitioners in design, engineering, electronics, sales, bookkeeping, accounting, and various levels of management. This type of ethical code contains excellent guidance urging employees to do the right thing, but unfortunately provides no guidance in assessing what the right thing is. While a code of this type is a guide for professional behavior, it does not provide guidance for recognizing and avoiding ethical dilemmas as they arise and cannot prepare the professionals to handle ethical dilemmas they may confront.

Additionally, codes of ethics are often developed by the legal staff of a firm. However, attorneys have a different standard of behavior, especially in the attorney–client privilege context, than that advocated in this book for practicing environmental leaders and professionals. Historically, maintaining confidential attorney–client communications is one of the highest duties of an attorney. An extreme example of application of this "attorney-specific ethics" component by attorneys happened in Southern California.[8] Two suspects were being defended against charges of murder and robbery, each by a different attorney. They were both in an interview with their respective attorneys when the first attorney received a phone call from the other attorney, who said, "My defendant just confessed to the crime but said it was your client that shot the victim." The first attorney repeated the information to his defendant, and the defendant opened his brief case and pulled out a gun. He said, "That rat, I'll show him what we do with squealers!" and jumped up and ran out of the attorney's office. The legal question was whether the first attorney should call the other to warn his client that he is under attack. The Bar Association's response was no! The information the attorney had was between him and the defendant and was confidential. Therefore, under the attorney–client confidential standard the attorney could not make that call. Subsequently, the Bar Association revised its finding and made an exception to the confidentiality rule allowing attorneys to take any action required to prevent physical harm. This is valuable in dealing with an environmental consultant's work performed under a client–attorney privilege contract. See Section 4.2.3, where a similar case has been discussed as an example ethical dilemma. Here if the consultant's finding discovers an imminent danger to public health, the professional's primary duty toward protecting public health clearly trumps adhering to attorney–client privilege components of the contract. On the other hand, the attorney involved will find that keeping the client's information confidential is higher on his priority of duties.

Often a firm has an ethics session once or twice a year during which employees are given a lecture on the company's code of ethics and then they sign a statement affirming they have read and understand that code. These sessions never really deal with the human instinct of self-preservation.

Environmental professionals consider the sessions perfunctory, and the presentations seldom have an impact on routine operations when they go amiss. It is insufficient for a firm to produce a code of ethics and have employees sign a certification indicating that they have read and understand that code, which only creates the illusion of protecting the managers in the organization from legal liabilities in case of employee error in the ethical discharge of their duties. A constant refresher in the form of a regular ethics workshop in which attendees are confronted with ethical dilemmas that an environmental professional might encounter on the job is needed. These could be one-hour sessions held at least monthly and can be conducted over a lunch provided by the host. Initially, a trainer should be used to establish a format, but after a few months the team can agree to a leader who will mediate the discussion. This book is designed to be a handbook for such a leader.

4.1.4 Participatory Workshops and Case Studies

There are many books on ethics that deal with the philosophy of human conduct under stress, and professors often lecture on the subject in hypothetical terms using, for example, the classic story of two people on a raft with only sufficient supplies for one person to survive. The generic study of ethics (i.e., doing the right thing) is interesting but so generalized it often cannot be applied in a real-world environment. When faced with a real-world ethical dilemma, a person cannot formulate an appropriate response from the hypothetical two-on-a-raft situation. In this book, over fifty realistic ethical challenges for environmental professionals are identified and analyzed.

Training is not a one-time activity but an ongoing process. The ethics training model we suggest for environmental professionals is a planned series of participatory workshops in which attendees review basic principles and are encouraged to discuss real-world situations they have encountered. There should be a significant number of participants to debate opposing opinions, thereby promoting deeper understanding. Each session begins with a facilitator introducing the six pillars of character (trustworthiness, respect, responsibility, justice and fairness, caring, and civic virtue and citizenship),[9] which participants draw upon and use for dealing with challenging issues with a particular focus on the environmental context. The facilitator may use the set of slides given on the CRC Press website to present introductory remarks for a company's ethics workshop. After some discussion, the remainder of the session is spent solving ethical dilemmas.

The case studies discussed in Chapters 5 and 6 of this book include examples of situations that environmental professionals may face in the field or office. By reviewing these examples with employees and by implementing their own set of ethical values, organizations will plant a seed of consciousness among their staff that will discourage them from wrongdoing.

Environmental professionals are often faced with issues that require them to use their judgment to ensure the actions they take are right. People do not always inherently react in the right way, and so must be trained to consider all the issues involved and openly discuss them with colleagues to foster a culture of ethics. While a culture of ethical behavior flows from the leadership downward, it also must be built ground up. It starts with educating students, training mid-level managers, and giving environmental professionals practice where the rubber meets the road. It is also important that people across the organization talk to one another and, by doing so, reinforce doing the right thing rather than conspiring to do the wrong thing. A key problem in the workplace is the reluctance to criticize colleagues for fear of antagonizing them. The need for acceptance can be a barrier to objective criticism. A formal ethics workshop with a planned agenda of discussion topics and probing questions can often pierce that barrier and help build momentum for learning to do the right thing. This participatory workshop model helps build an ethics culture in organizations.

4.2 Conducting an Environmental Ethics Training Program

Ethical behavior is learned and humans are born with an instinct for self-preservation. Ethics training is needed to combat this natural instinct. Once again, just as athletes must train their bodies to perform at their peak, environmental (and other) professionals require ethics training to ensure that they continue to behave ethically in difficult situations.

4.2.1 Structure of the Environmental Ethics Training Program

The most important parts of ethics training are periodic discussions of real or hypothetical case studies. Since discussions of this type seldom occur spontaneously, they must be planned and conducted on a regular basis. The first few sessions would cover basic values, the obligations of the environmental profession, management's position, and the firm's *modus operandi* for dealing with ethical issues. Subsequent sessions should concentrate on case studies—evaluating alternatives using the value system and *modus operandi* learned previously.

We recommend the ethics program follow appropriate organizational groupings. Preferably, the people in the group would be in the same area so that periodic one-hour sessions could be convened as easily as any routine meeting for communicating with employees. Ideally these discussions would be held during regular business hours, but sessions conducted over the lunch break are effective if employees willingly agree to participate in this manner. This arrangement works best if the employer provides lunch. The corporate ethics trainer would train a facilitator in each group, provide training material, and observe the individual workshops to help improve quality. Facilitator training consists of bringing together those who volunteer or are selected to facilitate and the trainer, and the selected trainees would conduct a workshop similar to those the facilitators would eventually conduct themselves. In addition to the information discussed in the previous paragraph, facilitators would learn how to conduct sessions without preaching to attendees, how to deal with strong individuals who might try to dominate sessions, and how to sensitively draw out timid participants. Ideally, the group manager and a representative from top management should appear at the initial sessions of the facilitator group and the individual groups. They should state the company's position and inform the employees of their duties and responsibilities regarding doing the right thing. When possible, a daylong seminar with top management should be held to acquaint them with the program, review the values being discussed, and reach a consensus on the *modus operandi* to be used when an ethical dilemma arises. The trainer can act as the facilitator in this top management meeting after presenting the introduction to the program. However, some thought might be given to using a third-party facilitator for the meeting, since the trainer will be somewhat of an advocate in this setting. If the firm is large, with many individual groups, the introductory presentation by top management can be accomplished by video, and the trainer would work with corporate staff to prepare a presentation outline and script. This video would be a valuable statement of the company's concern that all of its employees recognize the issues and do the right thing. The U.S. Federal Sentencing Guidelines recognize an effective environmental compliance program including values and ethics training.[10] In the event of an inadvertent environmental law infraction by an employee, this training video could provide convincing evidence of management's commitment to the environment. This in turn could persuade the U.S. Department of Justice or the Environmental Protection Agency to refrain from any criminal prosecution according to the guidelines.

Most organizations do not have a designated ethics official, although this trend is beginning to change. Therefore, any environmental professional supervisor should include a periodic ethics workshop in the training program. This section explains how to develop and conduct such an ethics training program.

Step 1: Conduct a meeting with your staff or the organization you wish to train. Explain your objective—to open a dialogue and train participants to be comfortable with reacting to ethical situations. Select an ethical situation they all know about, or use the classic one in which a client does not want you to reveal a violation you discovered. In the initial meeting, attempt to learn situations the group has experienced. They do not have to reveal how they handled that situation. Explain that the purpose of the following sessions is to collectively work out the best procedure for handling such ethical dilemmas. Finally, discuss the timing and duration of subsequent meetings.

Step 2: Use trained facilitators to conduct the first five workshops, and designate in-house staff to be trained as facilitators. The facilitator should make use of visuals such as PowerPoint slides to introduce each workshop for at least the first six months or year, depending on the frequency of the workshops. The facilitators should be outgoing and analytical and be bright and able to understand the issues and ask leading questions to stimulate discussion. Solutions should surface from the discussion, but the facilitator should be able to recognize the solution when it does arise. Finally, designate a recorder to document the discussion, though minutes should be kept confidential because revealing them could limit discussion.

4.2.2 Ethical Decision Making—Example Case Description[11]

It often takes training to even recognize an ethical dilemma. For example:

> You are a successful young project manager in a consulting firm that has national accounts with several major manufacturing firms. Because of your previous excellent performance, you are assigned to manage environmental assessments at several plants of one of these large companies. The contract is being conducted under legal counsel, and the results will be handled to maintain privilege under the law. You find a toxic chemical release that you know will dearly threaten public health. You contact the attorney who ordered the study and report your findings. The attorney directs you to stop work at that site, move on to another plant, and not to submit your findings in writing. You point out the need to report this to the responsible agency. The attorney reminds you of the contract that provides for the confidentiality of the findings and tells you that if there is to be any reporting, the attorneys will handle it. You are warned that this information and this conversation are to be held confidential and "will remain between the two of us. Do not discuss it with anyone!" Your attempts to learn how the attorney will proceed on this information are met with the comment, "It is not your concern. You do your job [i.e., investigate] and I will do mine. If I need any additional information, I will ask for it."

You do not know when or if the release will be reported. From the conversation with the attorney, you know that any attempt you make to learn further details on the disposition of your information will be met with hostility. Should you follow orders and go to the next plant? Should you document the findings in a formal report? The client might not pay for this action, since you were ordered not to prepare the report. Whether or not the law in this state requires you to report the release, if some adverse health effect occurs, the injured parties could sue the manufacturing firm. Eventually, it will be discovered that you and your consulting firm had knowledge of the release. What if your client decides not to report it?

This is truly an ethical dilemma and requires some difficult decision making. If you have never contemplated such a dilemma, you might decide to follow the attorney's dictates. After all, attorneys are supposed to tell you what to do under the law. Besides, your firm signed a contract that says the data are confidential. You might think, "OK, I will go to the next site and not make any waves."

If this scenario or one like it had been discussed in an ethics training session, you might have learned that your firm would advise that you elevate this situation to your supervisor, or further if necessary. Your management would want to ensure that some corrective action is taken to protect public health. Also in that training session, you would have learned that, as an environmental professional, your first obligation is to protect public health. You would know that to drop the issue and go on to the next plant is not the right thing to do. You would be prepared to do the right thing. The specifics of how this issue would be resolved in any consulting firm depend on the relationship of those involved. Some considerations will be discussed in the following paragraphs. But, if the consulting firm had held ethics training on a regular basis, the management's position on situations like these would be known and could be implemented with confidence and diplomacy.

4.2.3 Example Case Study Discussion[12]

Having established the responsibility of the environmental consultant to protect public health, we can explore how this responsibility applies in specific cases. Consider the case already described. The young project manager has just ended the initial phone conversation with the client contact, the attorney. If the consultant auditor has had ethics training, he or she would follow the values checklist (see Table 4.1) before elevating the situation to management. It is most important to be sure of the facts. Could there be any error in the data? Have verification tests been run? Should an uninvolved colleague be consulted to review the findings before anyone goes to management? Clearly, the confidentiality of the data must be maintained. However, the consultant cannot afford to be wrong when so much is at stake.

After all these issues have been checked and the necessary analyses conducted to indicate the certainty of risk to public health, before the call to the client, it is time to consult management. If this checking is still underway and will take some time, then management must be advised and informed of the verification plans. It is unacceptable for management to learn of the situation for the first time in a phone call from the client.

Ideally, a management team meeting will be held at which the project manager can lay out the facts. The communication link between the consulting firm and the client will be decided, usually at the highest level in each firm that has an established relationship. Necessary action should be taken to report the toxic pollutant release. The project manager should participate to the extent the client wishes but should keep aware of the proceedings at all times. At times like these, it is important to maintain respect and mutual concern. The client's attorney may seem dictatorial initially, and the project manager may feel some hostility. However, after the two company managements have made a plan of action, the project manager and attorney will probably need to work together. Therefore, it is important that the project manager maintain a respectful demeanor and an attitude of competence in dealing with the attorney from the outset.

Human nature can cause some of us to react spontaneously to difficult situations, thereby upsetting a business relationship. The situation described in the case study could cause a spontaneous reaction. However, the project manager's exposure to a hypothetical situation such as this one in an ethics workshop would be a reminder of the need to practice restraint and maintain dignity. In most situations a result satisfactory to all parties can be achieved. However, in the event that your management does not respond appropriately, you are not relieved of your responsibility to do the right thing.

You must pursue the proper action in a dignified manner, keeping in mind the values of justice and fairness. Try to understand the position of those with whom you disagree; try to convince them with facts and reason. Protecting the public from exposure to toxic pollutants is primary, but only if doing so does not create a more adverse impact. A more adverse impact could involve immediate physical and emotional stress caused by the loss of jobs when a plant is closed after severe penalties are imposed as the result of an inadvertent release by a confused employee. Compassion for families that might suffer due to plant closure is not inconsistent with doing the right thing. Each individual must establish his or her own values, which in some cases will mean leaving a firm because of a disagreement in values. Before one makes a decision of this magnitude, it is important to use a checklist such as the one shown in Table 4.1 to evaluate the issues involved.

4.2.4 Values Checklist for Example Case Study[13]

This values checklist (Table 4.1) requires some serious contemplation and reflects the thoughts of the authors. It is neither complete nor inherently correct. We suggest the reader use it as a point of departure to examine all sides of an issue, and this format be used whenever a professional is faced with an ethical dilemma. After consulting or completing the table, take your contemplations to a colleague, supervisor, or better yet, a group of confidants for a thorough discussion.

Ethics training is valuable for everyone. It sensitizes us to ethical issues and prepares us to respond appropriately to ethically questionable situations, which are usually unexpected. Workshops that include management participation are especially effective because management can share their values and employees can feel comfortable about their probable reactions

TABLE 4.1

Values Checklist for the Example Case[14]

Value	Issue	Action
Trustworthiness		
Honesty	Accurate and thorough reporting and interpretation of analytical data and answering all questions. Giving appropriate credit to participants in the report.	Disclose the findings, do not conceal questionable details, hold accurate but unrestrained dialogue. Involve your staff.
Integrity	The courage to report adverse findings, realize the possible consequences, and to advise clients in advance of this intent.	Report promptly to the client and your supervisor. (The case study does not indicate that the firm notified the client of its intent to report.)
Sincerity	Showing concern and appreciation for the seriousness of a situation and possible consequences.	Maintain a serious and concerned attitude in all dealings with the client.
Loyalty	Looking out for your firm's interest in a delicate position.	Inform your supervisors and accept this participation in the disposition.
Promise keeping	Meeting the explicit and implicit contractual requirements.	Maintain the confidentiality of the assessment data.
Respect		
Courteousness	Politeness and respect in all interactions with clients, supervisors, colleagues, and subordinates.	Say "please" and "thank you" and acknowledge clients' assistance.
Punctuality	Being on time for meetings and meeting program due dates.	On occasions when you may miss a due date, notify the client as soon as possible.
Right of self-determination	Respect for each individual's right to decide for oneself, even if wrong.	Provide clear direction initially and do not criticize others in public.
Responsibility		
Pursuit of excellence	Maintaining knowledge of the latest in technology and proficiency in the tools of your profession.	Ensure that the release detection reflects best measurement techniques.
Competence	Maintaining control of the situation from technical, safety, and management standpoints.	Demonstrate competence.

(Continued)

TABLE 4.1 (CONTINUED)

Values Checklist for the Example Case[14]

Value	Issue	Action
Integrity	The courage to do the right thing when the consequences may be adverse.	Ensure that the client (or responsible party) will execute cleanup as appropriate.
Self-restraint	Considering facts and circumstances. Consulting supervisors and colleagues before acting with care.	Elevate the situation to management and facilitate a plan of action.
Justice and Fairness		
Open-mindedness	Recognizing that there are various solutions to every dilemma, willingly considering them, and offering to cooperate with independent assessments when the seriousness of the matter warrants them.	Try to learn and understand any workable plans of the attorney. Suggest client get a second opinion to verify findings.
Ability to admit errors	Recognizing when you are wrong and openly admitting it. Releasing work product to be checked in an independent evaluation.	If there are any anomalies in the data, reveal them immediately. Get uninvolved colleagues to check work.
Caring		
Kindness	Helping others achieve their legitimate goals.	Offer to help the attorney deal with the situation. Explain risks and benefits.
Generosity	Offering money and free help and advice to a colleague for problem solving, etc.	Offer to meet with the client on an off-the-record basis to discuss issues.
Compassion	Recognizing the downsides of various actions for employees' families and other stakeholders and attempting to mitigate hardship.	Keep in mind the impact on stakeholders as you seek to resolve this dilemma.
Avoiding harm to others	Protecting public health and being concerned for public welfare.	Continue to seek prompt resolution to mitigate health and welfare impacts.

before dubious situations occur. Imagine a CEO who stands up in the workshop and says, "I have always respected the truth. You will never have trouble with me if you tell the client the truth and come directly to me with any conflicts." The employees in that workshop would have enhanced confidence in their employer's support for doing the right thing. In addition to reducing the risk of employees doing the wrong thing at the spur of the moment, these ethics workshops build confidence between management and employees that strengthens the relationship on all fronts.

We recommend semimonthly or monthly seminars during which a moral dilemma is discussed. These discussions should follow a half- to full-day workshop at which the values suggested here are reviewed and discussed, along with a presentation of management's recommended protocol for handling ethical issues.

4.3 Educating Students and Leaders in Environmental Ethics

4.3.1 Educating Graduate Students in Environmental Studies[15]

The subject of environmental ethics is in its nascent stage. While there are no extensive studies and data available on environmental ethics programs in U.S. graduate schools, sustainability programs could serve as a surrogate measure. Including sustainable engineering concepts into engineering programs is becoming widely accepted (80 percent of the top 100 engineering programs include them in some way).[16] A recent National Science Foundation (NSF)–funded study benchmarked programs and observed that the progress of sustainable engineering is at a critical juncture.[17] While there is significant grassroots activity, the path forward will require the development of a set of community standards for sustainable engineering.

Recognizing the need for such training, the academic Accreditation Board for Engineering and Technology (ABET) now requires ethics training as part of its board certification requirements for environmental professionals. To maintain professional certification, engineers must graduate in engineering from an ABET-accredited university, and in many states must take a specified amount of refresher training in ethics. The objective of this book is to provide guidance and text material to support university education in ethics as well as ethics workshops for training practitioners, leaders, and working groups in industry and in related consulting activities.

For the most part, graduate students in the environmental field, as in most fields of study, are not experienced in the various aspects of business. They may understand the concepts of right and wrong from their experiences in youth and college activities and occasionally from nonprofessional work experience. As discussed previously, they probably have some ideas that could lead them to make hasty decisions, intuitive decisions, decisions that they may not have made after giving the issue further consideration. They must learn to recognize ethical issues as they arise and how to appropriately respond to challenges they may face.

They may have learned the technique of programming and operating a computer. They may be skilled in public speaking and clear thinking. But ethics training is rarely practiced well and needs to be engrained in the students' mind before they face the challenges of the work world.

In the integrity recession section, we discuss how business and government leaders are attempting to cope with the current downturn in business, and how cheating and cover-ups are occurring as a natural outcome based on human instinct. We describe how managers have attempted to handle the situation by issuing codes of ethics detailed enough to cover specific activities. But the most effective training we have found is to identify improper actions and collectively decide how the dilemma should be resolved.

The previous section describes the process of acquainting graduate students with the nature and seriousness of these issues as well as our recommended approach to educating them. For the environmental graduate student, the problem often lies in recognizing an ethical dilemma so that it can be solved. We recommend dividing the class into groups of five to ten, with one student (or the teacher) acting as a facilitator. When a resolution has been reached, one member of the group will report the findings to the reassembled class. Once the reporter reads the finding, the reporter or teacher will moderate a discussion in which the other members of the class will offer their critique or approval of the findings.

The faculty advisor should oversee the small groups to ensure that participants are speaking appropriately and add any stimulating comments to initiate discussion. The teacher should take an active part in the group discussion and then summarize the findings for the students to record for their retention and future reference.

We recommend conducting a class session to explain the issues related to personal behavior and ethics. Then a later part of the class could be spent discussing ethical dilemmas in the manner described above. Finally, the instructor should encourage students to take this material to their workplace and suggest that regular ethics workshops be organized for ongoing refreshers.

4.3.2 Educating Business School Students[18]

Environmental issues are sociopolitical choices, and governments and corporations are facing a plethora of ever-expanding management challenges related to issues spanning physical, financial, geopolitical, and social concerns. With increasing engagement and scrutiny from the marketplace (regulators, investors, financiers, and consumers), integrated financial and sustainability reporting is shifting from voluntary to vital. Corporations and markets, including investors, are beginning to track and transform sustainability performance. The materiality of these seemingly noneconomic impacts is the critical link between organizational integrity and long-term value. To govern organizations, corporations, and leaders, new competencies in managing transparency, accountability, and in particular, an ethics culture must be constructed for business success in the next economy. Leaders need insight into how to determine which environmental, social, and governance metrics are material to them and relevant to their business and how best to manage them. Also, today young people actively seek careers that align with their values and give them the prospect to make a positive impact on the world through their work, which complements this need for education and training at the top.

Driven by these factors, sustainability and green business have swept through hundreds of management programs, both in the United States and around the world. Integration of ethics and sustainability education and training in business schools has been evolving over the past two decades. Several academic organizations, such as the Academy of Management's Organization and the Natural Environment (ONE), Alliance for Research in Corporate Sustainability (ARCS), and Association to Advance Collegiate Schools of Business (AACSB), promote sustainability-related research and teaching. Given the growing attention granted to companies that operate ethically and to meet the need for corporate leaders that can build an ethical culture, the Association to Advance Collegiate Schools of Business has recommended incorporating ethics content into business school programs.[19] The World Environment Center and Net Impact also affirm that graduates need to combine inside skills critical for understanding the company's internal functions with outside skills such as embracing the perspectives of customers and other stakeholders.[20] The organization aims to inspire as well as offer practical tools to business students interested in pursuing careers to achieve "a net impact that benefits not just profits, but people and planet too."[21] And of course the purpose of ethics, the dimension of sustainability.

Aspen Institute's biennial Beyond Grey Pinstripes ranks the global top 100 business programs that integrate social, ethical, and environmental issues into coursework and faculty research.[22] Hundreds of programs are including courses with ethical, social, and environmental content. Faculty who come from a variety of disciplinary backgrounds, such as strategy, operations, entrepreneurship, marketing, and finance, are integrating sustainability topics into the coursework in those disciplines. Many of the leading programs emphasize experiential learning through business cases, simulations, and consulting projects. It is quite telling that Harvard Business School Publishing, the premier source of case materials for business school classes, now boasts thousands of cases that feature companies tackling social and environmental challenges and opportunities in their businesses.[23] Some schools tend to emphasize the environmental aspects of sustainability, while others focus more on social attributes, typically under the banner of social enterprise. Few schools attempt to integrate the two.

Both faculty and industry representatives see a need for greater exposure to other disciplines that interact with and affect business, such as policy, science, technology, and complex systems.

Most MBA programs, however, do not have room to give students enough scientific literacy to deeply understand the technical depth of environmental issues companies face. To fill this void, a few institutions offer dual degree programs, in which students pursuing an MBA can simultaneously pursue a more science- or technical-oriented master's degree in environmental science or management.[24] These joint degree programs offer tremendous value, to both students and potential employers, as graduates are well versed in both the technical and business aspects of corporate sustainability programs.

Another challenge is that sustainability-themed course offerings are typically elective courses, chosen by a self-selected group, so the vast majority of business school graduates are not exposed to these topics. Integrating a full sustainability, ethics, or social responsibility course into the core business curriculum is unlikely in most places, so most schools take the easier path of introducing these concepts into existing courses. The benefit of this approach is that it reflects the real world, in which sustainability is more often embedded in all functional areas within an organization. However, the drawback is that there is not sufficient depth. Looking to the future, schools and programs that can balance the need for a depth of knowledge of sustainability as well as breadth across disciplines will produce the graduates most valued by employers. With increasing demand in the real world for professionals who can advance a new social contract between business and society, such graduates will be in greater demand, and will make a substantial impact on redefining capitalism with a purpose.

The Business Roundtable Institute for Corporate Ethics, established in 2004, recommends that ethics be a required core course, but given its interdisciplinary nature, it should be integrated into all business disciplines.[25] Starting in 2008, the Association to Advance Collegiate Schools of Business (AACSB) has held annual sustainability conferences, where faculty who have initiated innovative programs are given the opportunity to showcase their courses and teaching models, in order to share best practices in teaching both undergraduate business majors and MBAs.[26] In late 2009, the first Compliance and Ethics Forum brought together business and academic leaders to share leading thoughts and practices in the area of compliance and ethics.[27] Academic advisors for the Institute for Corporate Ethics guided the group in case-based exercises. This knowledge sharing was further enhanced by a presentation on a compliance leadership interactive training program. More recently, at the fourth ARCS summit in 2012, a panel of industry leaders, including several chief sustainability officers from Fortune 100 companies, was invited to offer critiques of how business schools are or are not meeting their needs.[28] The panelists commented that business schools were doing a fairly effective job of imparting technical-oriented business skills (e.g., discounted cash flow analysis and optimization). They further suggested that business schools need to place greater emphasis on soft skills, such as understanding how sustainability can be embedded into organizational culture and knowing how to negotiate tough trade-offs when environmental choices do not offer win-win solutions.

Baker introduced an experiential exercise for students of business ethics, Business Ethics Everywhere (BEE).[29] The coauthors observe that most of the current pedagogic tools focus on identifying ethical issues and applying philosophical ethical or moral frameworks or some other structured decision criteria, or both. While they increase students' ability to recognize and resolve ethical dilemmas, this is only a precursor to and a necessary element of possible ethical behavior. It does not prepare the student to implement these in the organizational context. They lack the pragmatic hands-on element of thinking through the implications of the chosen response for themselves and other impacted stakeholders. Business Ethics Everywhere (BEE) is presented as an experiential exercise taking them to the next step of how to tackle these ethical dilemmas. Students are asked to reflect on their recent organizational experience of ethical issues and consider what could have been done differently and why. The stated learning objectives for this exercise start with identifying the variety of workplace ethical issues, applying the framework of reasoning, and enhancing students' ability to make decisions that take into account the realities of the organization.

Our subject is perhaps more focused on environmental ethics. The attempt is to train environmental professionals by identifying myriad ethical dilemmas that the professional could face and offer suggested approaches for handling the dilemmas. This book offers a robust set of over 50 well-developed potential dilemmas one may encounter in the environmental ethics area in an organizational context. Before diving into the dilemma resolution, we discuss the rationale that the professional should follow in handling the situation. Finally, the suggested approach is to use the cases in frequent scheduled participatory workshops, which brings the learning objectives to fruition very effectively, although admittedly for the narrower range of ethics issues, namely, those likely to be faced by the environmental practitioner—leaders and professionals alike.

4.3.3 Ethical Leadership Qualities and Mechanisms

Developing organizational morals and creating morals for others has been identified as a leadership function in academic literature for over 70 years.[30] Vin Sarni, former CEO of PPG, says CEO stands for chief ethics officer.[31] Schein[32] describes three primary leadership mechanisms that leaders employ: model the role and demonstrate the behavior, institute an incentive or compensation system, and use ethics as a factor to hire (including promote) and fire employees. What leaders want employees to focus on matters.[33] More importantly, what they do matters even more. This has been very well said by Kent Druyvesteyn, former VP of ethics at General Dynamics:

> We could have Jesus and Moses and Mohammad and Buddha (and Krishna and Gandhi and Plato and Aristotle) come and speak at our (ethics) workshops. But, if after that one leader violated ethical standards; that teaches employees more than all the experts in the world.[34]

The expected behavior of ethical leaders to successfully serve as role models is summarized as the following:

1. Committed to high standards that apply to all
2. Uses what's best for public and consumer as a primary standard
3. Willing to confront honestly and openly
4. Takes timely and appropriate actions
5. Proactively heads off problems
6. Sets personal example and treats it as a priority
7. Treats it as a performance measure just like sales and profits
8. Learns about legal and ethical aspects of business, does not plead ignorance
9. Does not use lower standards of peers as an excuse[35]

But even "good managers often make unethical decisions—and don't even know it." Bias is identified as the major source of unintentional unethical decisions.[36] These include implicit prejudice that emerges from unconscious beliefs, in-group favoritism that favors your group, overclaiming credit that favors you, and conflict of interest that favors those who can benefit you. Authors caution that simple conviction, sincere intention, or trying harder is not enough.

The discussion above has focused on ethical leadership. As discussed in earlier sections, we see ethics (also referred to or incorporated in the term *governance* these days) as the fourth dimension extension of the triple bottom line commonly used in the context of sustainability. Hence, a professional leading environmental ethics issues should combine both sets of leadership qualities and mechanisms.

Visser[37] defines a sustainability leader as "someone who inspires and supports action towards a better world." This is particularly relevant to the environmental leader building an ethical culture. He identifies a set of qualities that include traits, styles, skills, knowledge, and characteristics as well as the internal and external actions expected of leaders.

1. Traits: The leader should be caring, open minded, empathetic, and courageous, as well as a holistic thinker.
2. Styles: The leader's operating style could be a combination of inclusive, visionary, creative, altruistic, as well as radical.
3. Skills: The expected abilities include managing complexity, communicating vision, exercising judgment, challenging and innovating, and thinking long-term.
4. Knowledge: A broad array is called for that comprises global challenges and dilemmas, interdisciplinary connectedness, change

dynamics and options, organizational influences and impacts, and an understanding of diverse stakeholder views.

5. Internal actions required of a good sustainability leader are making informed decisions, providing strategic direction, crafting management incentives, ensuring performance accountability, empowering people, and embedding learning and innovation.
6. External (stakeholder-related) actions encompass fostering cross-sector partnerships, creating sustainable products and services, promoting sustainability awareness, transforming context, and ensuring transparency.

In summary, key characteristics that are essential for a sustainability leader are systemic understanding, emotional intelligence, and values orientation, with a compelling vision, inclusive style, innovative approach, and long-term perspective.

Endnotes

1. Ethics Resource Center, "2004 Pace Leadership in Ethics Award Acceptance Speech by Norman Augustine," http://www.ethics.org.
2. Ethics Resource Center, "ERC Honors the Late Carol Marshall with the 2009 Pace Award," last modified March 25, 2010, http://www.ethics.org/ethics-today/0310/carol-marshall.html.
3. M.R. Banaji, M.H. Bazerman, and D. Chugh, "How (Un)ethical Are You?" *Harvard Business Review*, December 2003, pp. 3–10.
4. The Air and Waste Management Association (A&WMA) is headquartered in Pittsburgh, Pennsylvania, and its website is http://www.awma.org.
5. *EM, Environmental Manager*, is a magazine published by the A&WMA.
6. See appendix for the A&WMA's code of ethics.
7. AECOM, formerly ERT, is one of the largest environmental consulting companies in the world.
8. San Diego County Bar Association, "Ethics Opinion 1990–1," accessed December 2012, http://www.sdcba.org/index.cfm?Pg=ethicsopinion90-1.
9. Michael Josephson, "Making Ethical Decisions," accessed December 2012, available at http://www.sfjohnson.com/acad/ethics/Making-Ethical-Decisions.pdf.
10. Reynolds Wilcox and Louise Theodore, "U.S. Federal Sentencing Guidelines and the Development of Ethics Education Programs in the Environmental Industry" (paper 96-TP16IA.05 presented at the 89th Annual Air and Waste Management Meeting, Nashville, June 23–28, 1996), p. 273.
11. Hal Taback, "Ethics Training: An American Solution for Doing the Right Thing," in *Engineering and Environmental Ethics*, edited by John Wilcox and Louis Theodore (New York: John Wiley Price, 1998), pp. 267–274.

12. Ibid., pp. 267–274.
13. Ibid., pp. 274–279.
14. Ibid.
15. Ram Ramanan, "Cover Story—State of Post Graduate Environmental Education in the US and the World," *Environmental Manager*, September 2012, pp. 1–5; H. Taback and R. Ramanan, "The State of Ethics Training for Environmental Professionals," *Environmental Manager*, September 2012, pp. 25–27.
16. David T. Allen, "Incorporating Sustainability into Engineering and Teaching" (presented at A&WMA Annual Conference and Exhibition, San Antonio, TX, 2012).
17. Cynthia S. Murphy, et al. "Sustainability in Engineering Education and Research in US Universities," *Environmental Science and Technology* 43, no. 15 (2009).
18. Ram Ramanan and W. Ashton, "Green MBA and Integrating Sustainability in Business Education," Air and Waste Management Association's Environmental Manager, September 2012, pp. 13–15. Accessed December 2012, http://stuart.iit.edu/about/faculty/pdf/green_mba.pdf.
19. Ethisphere, "World's Most Ethical Companies," accessed December 2012, http://ethisphere.com/worlds-most-ethical-companies-rankings/.
20. World Environment Center and Net Impact, "Business Skills for a Changing World," accessed December 2012, http://www.wec.org/events/net-impact-wec-business-changing-world.
21. Ibid.
22. Aspen Institute Center for Business Education, "Beyond Grey Pinstripes," accessed December 2012, http://www.aspencbe.org/.
23. Harvard Business School Publishing, "Insights and Publications," accessed December 2012, http://harvardbusiness.org/.
24. These schools are American University, Duke, Illinois Institute of Technology, University of Michigan, and Yale.
25. An association of CEOs of U.S. companies with $6 trillion in annual revenues and 14 million employees, housed at the University of Virginia's Darden School of Business, which links ethical behavior and business practice. See http://www.corporate-ethics.org; *Shaping Tomorrow's Business Leaders: Principles and Practices for a Model Business Ethics Program* (Charlottesville, VA: Business Roundtable Institute for Corporate Ethics, 2007).
26. AACSB, "The Association for Advancement of Collegiate Schools of Business (AACSB) Sustainability Conference," accessed December 2012, http://www.aacsb.edu/sustainability/.
27. *Compliance and Ethics Forum, Summary Report, Leading Thoughts and Practices* (Charlottesville, VA: Business Roundtable, Institute for Corporate Ethics, 2009).
28. "Industry Leaders Panel Discussion" (Fourth Annual Conference of Alliance for Research in Corporate Sustainability, Yale University, CT, May 16–18, 2012), accessed December 2012, http://www.corporate-sustainability.org/conferences/fourth-annual-research-conference/.
29. Susan D. Baker and Debra R. Comer, "Business Ethics Everywhere: An Experiential Exercise to Develop Students' Ability to Identify and Respond to Ethical Issues," *Business Journal of Management Education* 36 (2011): 95–125.
30. C.I. Barnard, *The Functions of the Executive* (Cambridge, MA: Harvard University Press, 1938), p. 272.
31. L.K. Trevino and K.A. Nelson, *Managing Business Ethics: Straight Talk about How to Do It Right*, 2nd ed. (New York: John Wiley & Sons, 1999).

32. E. Schein, *Organizational Culture and Leadership* (San Francisco: Jossey Bass, 1985).
33. E. Schein, *Organizational Culture and Leadership* (San Francisco: Jossey Bass, 1985); L.K. Trevino and K.A. Nelson, *Managing Business Ethics: Straight Talk about How to Do It Right*, 2nd ed. (New York: John Wiley & Sons, 1999).
34. L.K. Trevino and K.A. Nelson, *Managing Business Ethics: Straight Talk about How to Do It Right*, 2nd ed. (New York: John Wiley & Sons, 1999).
35. Ronald R. Sims and Johannes Brinkmann, "Leaders as Moral Role Models: The Case of John Gutfreund at Salomon Brothers," *Journal of Business Ethics* 35 (2002): 327–329.
36. M.R. Banaji, M.H. Bazerman, and D. Chugh, "How (Un)ethical Are You?" *Harvard Business Review*, December 2003, pp. 3–10; Scott A. Quatro and Ronald R. Sims, *Executive Ethics—Ethical Dilemmas and Challenges for the C-Suite* (Charlotte, NC: Information Age Publishing, 2008), pp. 3–10.
37. Wayne Visser and Polly Courtice, "Sustainability Leadership—Linking Theory and Practice," SSRN Working Paper Series, October 21, 2011, accessed December 2012, http://www.waynevisser.com/wp-content/uploads/2012/06/paper_sustainability_leadership_wvisser.pdf; also presented in the Cambridge Program for Sustainability Leadership (CPSL) (2011) report entitled "A Journey of a Thousand Miles: The State of Sustainability Leadership 2011," http://www.cpsl.cam.ac.uk.

Section II

Environmental Ethical Dilemmas—Case Studies

This casebook on ethics offers to help the environmental professional appreciate the choices that others make and to evaluate the justification they give for those decisions. It also sharpens one's ethical awareness; that is, it invites the reader to undergo a conscious examination of values and choices and how these may have shaped his or her life so far, and how they can be used to shape it in the future. However, this book is not intended to outline theories of ethics and the terms used in ethical debate, but rather to examine the conflicts experienced by environmental professionals using actual dilemmas and solutions thereof. Some ethical choices are clear-cut; others are more problematic, and it is by looking into the problematic ones that we can start to appreciate the issues involved in resolving those dilemmas.

Ideally, this material would be used by university professors and organizational environmental managers and leaders for educating students and training employees who are performing environmental duties for developing and building an ethical culture. Additionally, groups of environmental professionals or individuals would use it on a scheduled periodic basis to maintain individual awareness of the ethical dilemmas that they may encounter in their practice and for maintaining their professional engineer (PE) and other board certifications. These participatory training workshops to help our professionals hone their skills and prepare our students form theirs to make ethical decisions may be one significant step to counter the myriad temptations and opportunities to stray that entice all of us in today's business world.

Chapters 5 and 6 of this book identify and deal with numerous dilemmas in a case study format, offering various options, testing ethical values, and giving trainees practice in the challenge of doing the right thing. Chapter 5 focuses on the job conflict situations faced by practitioners. Chapter 6, on the other hand, addresses situations encountered that comprise a judgment or a decision that involves choosing one among multiple alternatives. The environmental case studies in this book are not all true ethical dilemmas, in which the right thing to do is unclear. Far more frequently, the dilemmas involve situations where the right thing to do is quite apparent, but difficult to fulfill. In these kinds of situations, it is important not to react too quickly, because a gut reaction can often get you into trouble. Instead, one should handle the situation by considering the potential impact on all the stakeholders involved and putting a set of principles to guide decision making into practice.

A large proportion of environmental professionals primarily carry out environmental compliance activities. They are environmental scientists and engineers providing surveillance and guidance for compliance with governmental regulations. Many other environmental professionals may opt to serve society in government planning and nongovernmental watchdog roles requiring similar expertise. In effect, environmental professionals may be part of the regulated community of company in-house professionals, professional firms or independent practicing consultants and attorneys, and government (federal, state, or local) regulators or nongovernment advocates.

In-house professionals in operating facilities learn and comply with various national, state, and local environmental protection agency regulations. They must comply with all the rules, such as getting permits, limiting emissions, measuring, and reporting. They must perform measurements confirming the plant's compliance with requirements and maintain records of these activities subject to review by various agencies. They must actively survey plant operations to ensure the compliance by plant personnel as they execute those requirements. Environmental professionals, like many others working in corporate roles, face both time and cost pressures. Routinely budgets and staff are reduced while the regulatory requirements show no sign of relenting or going down. Ways of working smarter and leveraging technology will reach their limits soon, especially in the short term. This leads to pressure from supervisors to reduce expenses, eliminate positions, and delay investment in process or operational improvements and changes that do not have immediate financial returns. These resource pressures lead to potential ethical dilemmas on the path to find ways to satisfy the supervisor's requests/demands. These may involve the following

1. Delaying valuable maintenance expenses that could have kept the plant operating in an environmentally efficient manner
2. Delaying emission reduction projects that could reduce releases and discharges below the mandated levels to further enhance protection of public health or reduce environmental impact

3. Overloading a well-performing employee and setting the employee up for failure or downgrading an employee's performance to meet company HR policy protocol before firing the employee

Professional firms of consultants and attorneys may be contracted to perform some of these compliance activities, depending on the size of the plant, the skills and experiences of the in-house professionals, and the frequency of the required action. The top management of a firm planning an expansion or executing major construction work for a client will consult with specialists to plan their moves. When applying for new equipment permits, professionals often must assess and model emissions from a new source and perform health risk assessments. On other occasions the regulatory agency may have issued a notice of violation or the company may have agreed to a consent decree to rectify the infractions. These transactions involve legal oversight and contracts and interactions among multiple parties.

1. Employees of consulting firms have to be especially alert to attempts to corrupt their findings.
2. On the other hand, supervisors in consulting firms have to be aware of any requests from clients that their staff members cheat. The idea that the customer is always right can get a consulting firm into trouble.
3. Law firms and in-house legal advisors bring a completely new level of complexity with attorney-client privilege ethical standards.

Finally, the government and nongovernmental roles require making decisions on prioritizing allocation of resources, identifying levels for controls needed, and establishing rules and regulations and audit protocols and associated incentive and deterrent mechanisms. Also, because of their special qualifications, environmental professionals working in companies are encouraged by their employers to reach out and serve the community in which they live. By serving on a city or county board of supervisors, they can often use their specialized knowledge to provide guidance for community projects and help solve domestic issues that involve the environment.

1. Just by their inherent nature, these roles involve fiduciary duties on the part of every professional to make balanced ethical choices. These roles will encounter a multitude of group interests and agendas and have to contend with representations and lobbying.
2. Environmental ethical dilemmas arise at every permit approval, each public hearing on rule development, and hundreds of advocacy and activism demands, and clearly call for making ethical decisions. These professionals need to be ready every day to face these situations and need thorough training and ongoing reminders.

3. Regulatory agency inspectors and company in-house auditors face similar challenges on determining how strict they would be in interpretations of applicability and standards and how vigilant they have to be to unearth violations and resist pressures to not issue notice of violations. Their duties get harder in finding missed and deliberately hidden company internal reports.

5

Environmental Dilemmas—On-the-Job Conflict Situations

5.1 Client Attorney Directs Consultant to Not Report Findings

5.1.1 Dilemma

You are a successful project manager at Early Environmental Advice (EEA), a consulting firm with national accounts with several major manufacturers. You are assigned to manage environmental compliance audits at several plants of Crown Industries, Inc. (CII), a large company. The contract is being conducted under legal counsel and the results are to be handled in order to maintain confidentiality under the law. At one plant, you discover the release of a toxic chemical that you are certain threatens public health. You contact the CII attorney, Paul, who ordered the study and report your findings. Paul directs you to stop work at that site, move on to another plant, and not submit your findings in writing. You point out the need to report this to the responsible agency, Paul reminds you of the contract, which provides for the confidentiality of the findings. You are told that his staff will handle any reporting and warned that this information and your conversations are to be kept confidential. "Do not discuss it with anyone! If I need any additional information, I will ask for it. Take your team and go on to the next plant on the list." Your attempts to learn how Paul plans to proceed on this information are met with, "It is not your concern. You do your job [i.e., investigate] and I will do mine. If I need any additional information, I will ask for it."

5.1.2 Discussion

You have no knowledge as to when—or if—the toxic chemical release will be reported. From your conversations with Paul, you know that any attempt on your part to learn further details on the disposition of the information will be met with resistance. Should you follow orders and go on to the next plant? Or should you document the findings in a formal report? (Note that Paul will not pay for this action since you were ordered not to prepare it.) Regardless of whether or not the state law requires you to report the toxic chemical release, if an adverse health effect results from it, the injured parties could sue CII. Eventually, it will be discovered that the EEA staff was in possession of knowledge of the release and your firm could be included in the lawsuit. But what if CII decides to not report it? This is an ethical dilemma and requires some difficult decision making. If you've never contemplated such a dilemma, you might decide to follow the dictates of the attorney. After all, attorneys are supposed to tell you what to do under the law. Besides, EEA signed a contract that says the data are confidential. You might reason, "OK, I will go the next site and not make any problems." But should you? It is apparent that Paul does not trust you to keep your findings secret. The contract clearly calls for this secrecy. Besides this, the values of trustworthiness and respect require that you comply with the attorney's direction. On the other hand, the priority of protecting public health is our concern. Will Paul make the necessary notification to the proper agency? Why is he so reluctant to tell you his intentions for handling this situation? What if his intention is to correct the condition without any declaration? Clearly the right thing is to declare the toxic chemical release and take the necessary cleanup actions. But will they? Do you need to know?

5.1.3 Recommended Action

After speaking with the attorney, Paul, you should realize that an ethical dilemma exists. First, you should run verification tests on the data. Next, you should consult an uninvolved colleague to review the findings before reporting the problem to management. Do not reveal the conflict with the attorney to your colleague. Clearly, the confidentiality of the data must be maintained, but there is no margin for error when so much is at stake.

With verification completed and checked, you should elevate the issue to your supervisors. EEA should want to ensure that some corrective action is taken to protect public health, which is its first obligation. Just going on to the next plant is not the right thing to do. The manner in which this issue would be resolved by a given consulting firm depends on the relationship of the firm's principals with its communication skills. It must be handled with confidence, diplomacy, and integrity. The best approach is for a senior member of EEA (a corporate attorney, if one exists) to make a contact with a senior executive of the client. In most instances, the higher one goes up within the management organization, the more concerned the executives are likely to be. You should participate to the extent that the executive wishes. The necessary action should be taken to ensure that the toxic chemical release is reported.

At times like these, it is important to maintain respect and civility in all your dealings. CII's attorney seemed dictatorial initially, and you may have felt some hostility. However, after the two firms have agreed to a plan of action, you and the attorney will be working together. Therefore, it is important that you maintain a professional attitude in dealing with the client's attorney from the outset. As discussed earlier, human nature can cause some people to react spontaneously to situations, upsetting a business relationship. This kind of reaction must be avoided. In the event that the EEA management does not respond appropriately, you are not relieved of your responsibility to do the right thing. You must pursue the proper action in a dignified manner, keeping in mind the values of justice and fairness. Try to understand the position of those with whom you disagree. Try to convince them with facts and reason. Protecting the public from exposure to toxic pollutants is primary, but only if a more adverse impact is not created as a direct result of some action. Such a possibility could involve immediate physical and emotional stress caused by the loss of jobs when a plant is closed due to severe penalties imposed on the facility as the result of an inadvertent release by a confused employee. Compassion for families that might suffer due to plant closure is not inconsistent with doing the right thing. Each person must establish his or her own interpretation of values, which in some cases may dictate leaving an employer because of a disagreement in values. In making a decision of this magnitude, it is important to carefully evaluate the issues involved.

5.2 Plant Operates Substantially beyond Permit Levels

5.2.1 Dilemma

You are an environmental consultant for Quality Environmental Services (QES), and your assignment is to perform a regulatory compliance audit at the plant of a national corporation, Utica Nautical Suppliers (UNS), an important client to your firm. After talking to plant personnel and reviewing production records, you determine that the UNS plant operates substantially beyond its permit limits, which could cause as much as 50 to 75 percent more pollutant emissions to be released than is allowed by its permit. However, official UNS reports submitted to the regulatory agency show that plant's emissions fall within the allowable limits. You inquire further and are told that the corporation was about to close the plant until the UNS plant manager installed some "efficiency improvements" to increase production. This saved the plant from closure. It also saved the jobs of 300 people in the small town where the plant is located and is the major employer. Because the plant is downwind of a large city, the area is in nonattainment. It would likely cost well over $1 million to purchase offsets to cover the resulting increased emissions. Installing new emission controls would cost even more. The UNS plant manager, George, prevails on you to keep his secret and save the jobs of his loyal employees. "Look," he says, "my plant does not create a health hazard in the community. If this plant did not exist, the air quality would be the same. The pollution comes in from the big city." He tells you that only three of his staff members know of this discrepancy. He also tells you that he is a personal friend of the chief of the local air pollution agency, who is not aware of this situation. What should you do?

5.2.2 Discussion

This dilemma looks at the ethical responsibility of an environmental consultant performing a regulatory compliance audit at an important client's economically challenged plant. You do not disagree with the issues noted by George. You understand that the plant is not a true danger to public health even in this noncompliant operating mode. Had you yourself not been so diligent, you might not have discovered the discrepancy. Nevertheless, the plant is violating the law. Closing the plant would affect 300 employees and thousands more in the local community. If the UNS corporate management is bottom line oriented, it will most likely close the plant, rather than pay large sums for emission offsets or control devices. Do you help the plant manager and keep his secret?

What do you think? The answer is yes and no: you try to help George, but you do not keep his secret. First and foremost, you have a responsibility to inform your QES superiors of the situation. In a situation like this one, you need all of the clear thinking that can be brought to bear on the situation. Do not try to go it alone. The values discussed previously should be considered: trustworthiness, respect, responsibility, justice and fairness, civic virtue and citizenship, and caring. Trustworthiness mandates that you do not keep George's secret, but tell the truth. Respect requires that you respect the plant manager's concerns for his employees and try to help him keep his plant running in the face of what he views as an unfair obstacle. Responsibility necessitates that you obey the law and convince George to comply with it. Justice and fairness calls for an understanding that, in this situation, the law may be unfair to the community. Civic virtue and citizenship mandates that when you recognize an injustice in the law, you do not ignore it. It is your civic duty to help correct it. Finally, caring requires that you care about the effect this issue will have on the employees and other citizens who are the stakeholders in this case. You should do what is ethical to help them, but without breaking the law.

5.2.3 Recommended Action

First, the person in QES with the closest professional relationship with UNS's corporate management should accompany you when you explain the situation to the client. Second, support UNS's efforts to obtain relief from new source requirements. If you know of a means to control the plant's emissions economically, suggest a design study. If not, offer to support the client in an effort to change the law by demonstrating to the agencies, legislatures, or even Congress that the current law is unfair. Encourage UNS to rally the local community to write letters to high officials, explaining the issue and suggesting actions to resolve it. This is clearly an unusual approach and should be taken cautiously. Third, suggest that UNS's attorney contact the local regulatory agency to request a variance until the matter is resolved. Do not hesitate to make use of the George's personal relationship with the chief of the local agency. Apparently, the UNS plant is old, and may be outdated compared to other company plants. If the company does not want to take such heroic measures to keep the plant open, then all you can do is hope that it will "do the right thing" to either update the plant or transfer the plant personnel to another location. The latter scenario will not help the ancillary businesses in the town, but that is not a matter that you are in a position to deal with.

(This is an issue of real concern because a small California town downwind of Los Angeles had a "severe nonattainment" classification due to pollution transport from Los Angeles. Due to this, a manufacturing plant in that town faced a limited-growth issue. The company asked the authors, "How can the EPA limit our emissions? If this whole town disappeared, the air quality would not change in this area.")

5.3 Boss Purposely Overestimates Client's Price for a Contract Task

5.3.1 Dilemma

You are an experienced test engineer with ten years of service at Warm Fuzzy Environments (WFE), environmental consultants. WFE specializes in indoor air quality, especially sick building syndrome. Your supervisor Jack, one of WFE's owners, assigns you to prepare a test plan for a client's 10-story office building. Jack is very detailed in the measurements he wants the test crew to take. You believe that the number of tests Jack has specified is much more extensive than is necessary. From further conversation with him, you are certain that your supervisor is just padding the workload to keep the WFE employees busy during a slow business period. Jack rejects your protests and implies that if you do not follow directions, you'll be the first one to be fired. You have just moved in to a new house with a sizable mortgage payment, and since your wife is expecting a baby, eventually one of you will have to stop working. How do you proceed?

5.3.2 Discussion

This dilemma is an example of one of the seemingly mundane issues that might be ignored or dismissed as a routine business practice. But it is something that needs to be discussed and addressed in any consulting firm. It certainly seems like an issue that can be handled simply by doing what the boss wants. If the client is disturbed by WFE's estimate, it can go elsewhere. And if it is so dependent on WFE that it does not even seek competitive bids, maybe it deserved to be taken a little. After all, there have been jobs for which WFE has had to absorb overrun costs. Besides, with your increased personal cash crunch due to the new house and baby, this is no time to risk your job. You might dismiss this as not an ethical issue at all. Maybe the readers would agree with that position.

The assumption here is that you are a competent practitioner in the field, and Jack has clearly included tasks that are not required to diagnose the condition of the office building. The A&WMA's code of ethics (appendix) has relevant items that apply to this situation. Item 7 states, "Exercise honesty, objectivity, and diligence in the performance of all my professional duties and responsibilities." Item 9 states, "Act as a faithful agent or trustee in business or professional matters, provided such actions conform to other parts of this code." Finally, item 13 states, "Treat co-workers, colleagues and associates with respect and respect their privacy." You want to do the right thing, but without jeopardizing your job if possible.

5.3.3 Recommended Action

The first thing to do is to work with one or more colleagues to prepare a clear statement of test objectives, as you know them. After ten years, this should be straightforward. Then, prepare a test program that will accomplish those objectives. Explain why the additional tasks suggested by Jack are redundant or should be considered later, depending on the results of the basic tests that you have prescribed. Develop an estimate for those additional tasks. Estimate the schedule and cost of the two task categories. Prepare this information in a concise memorandum crediting the assistance of your colleagues. It might be necessary to do as much of this as possible on your own time in order not to antagonize your supervisor. Since your colleagues may or may not be willing to contribute their time, you can use them primarily to critique your proposal. It is important not to reveal your conflict with Jack to the colleagues in order to maintain your loyalty to WFE and not degrade the respect of your colleagues for Jack. It is clearly a delicate task to resolve this dilemma while minimizing the impacts on yourself and the other stakeholders (i.e., WFE and its employees).

Arrange for a private meeting with Jack. Present the material in sequence. Make sure that it is clear that the proposal is a joint effort between you and your colleagues. The fact that the others are involved should help you gain a more convincing argument for the supervisor to consider. Get agreement on the statement of test objectives. Modify it as necessary with your mutual agreement. Then review your test plan explaining how it will achieve the objectives. Then discuss how the additional tasks are redundant but might be considered as future scope if the basic tests are inconclusive or need further confirmation by the client. Maintain a level of respect, being certain not to allow any of Jack's remarks to cause you to react hostilely. Hopefully Jack will respect your efforts and will reconsider the earlier threats.

In anticipation of further discussion, you should be prepared to explain why you were concerned about Jack's earlier arbitrary direction and veiled threat. Review the association's code of ethics and be prepared to show them to Jack. Explain that you are cost-sensitive and believe that client loyalty is best served by providing diligent, cost-effective services. Do not use derogatory terms like *padding*. Maintain respect at all times. If your supervisor will not listen and reacts hostilely to you, then take your case to the next level of management.

Jack may respect this approach, which could possibly cause him to reveal that he is under pressure from the top management to maintain or increase revenue. You might discuss how doing the right thing here would be beneficial to the health of the firm. If he is receptive, have an informal chat with Jack to evolve a marketing plan to present to management. If this fails and you are not comfortable with this negotiation, back off and proceed as necessary to maintain your job in view of the domestic demands on you. However, you should start investigating other employment opportunities. WFE is no longer a company with which you should be associated.

5.4 Contractor Holds Back Completed Report for More Money

5.4.1 Dilemma

You are the program manager of a comprehensive state agency-sponsored two-year regional air quality study. There is one prime contractor, a public research institute, that employs multiple subcontractors. The prime contractor was selected after a proposal competition (you did not participate in this contractor's selection) and accepted a fixed-price contract to conduct the program. As the project nears completion of the fieldwork phase and enters the critical phase of data analysis and report preparation, the contractor begins to miss certain project deadlines and periodic reports. You meet with the contractor's principal investigator to determine the cause of the delay and explore ways to improve the schedule. The principal investigator offers excuses and promises to meet all future deadlines. However, more deadlines for completion of work are missed, resulting in more excuses and promises, but no improvement. You determine that the technical work done so far by the contractor has been performed competently. Nevertheless, owing to the importance of the project and the many participants involved, the periodic reports are critical.

A few months pass by and a critical deadline approaches—delivery of the draft final report. Many parties are counting on receiving this report on time. The deadline comes and goes, but the contractor fails to deliver the report. When you contact the principal investigator, you learn that the report is complete, but that the contractor wants more money (between 5 and 10 percent of the contract's original value) to deliver the report. The principal investigator is a tenured scientist with the public research institute and he tells you, as an apparent threat, that the institute is not concerned about being sued. However, those issues are secondary, as you are under great pressure from many parties to deliver the report, and on time. How do you proceed?

5.4.2 Discussion

The principal investigator violated five of the six pillars of character: trustworthiness, respect, responsibility, fairness, and caring. Trustworthiness includes honoring commitments of time and money. It would be an egregious violation of ethics to commit to performing a task in a proposal where price is an element for selection and then withhold the product at the last minute demanding more money. In addition to promise keeping, trustworthiness involves honesty and integrity. It was dishonest of the principal investigator to mislead you about the late reports. Respect concerns courtesy, civility, and decency. Therefore, how can an environmental professional treat a client with such disdain? Responsibility includes accountability, self-restraint, and the pursuit of excellence. The granting of academic tenure does not relieve one of responsibility. The issue of fairness was violated when the principal investigator said that the schedule would be recovered. Even if the principal investigator believed this to be true at the time, he had a duty to employ a process that would ensure that he learned the truth before making promises he couldn't keep. Finally, the principal investigator showed no caring in causing embarrassment to the project manager. Similarly, he risked damaging the institution's reputation.

Having established that the principal investigator's actions were contrary to core ethical values, how should the project manager proceed? Litigation is the obvious initial reaction. However, if the state agency sued the contractor, the goal of the study could be delayed for a long time and eventually defeat the program. Assuming that the contractor was a federal or state research laboratory that is funded on a project basis by other governmental agencies, bringing a lawsuit against a governmental entity would have enormous political ramifications. As project manager, you must handle the situation in the best interests of all stakeholders.

5.4.3 Recommended Action

Since many parties are depending on receiving the final report, it is reasonable to assume that it is a large project with a commensurate budget, of which the prime contractor's research effort is a significant portion. It was specified that the research performed by the contractor was competent. In the course of your interactions with the principal investigator, you should have been able to assess the progress, and expected that the final report would provide the data and interpretation required by the study. The problem lies in obtaining the report.

Assuming that your agency has paid all of the contractor's outstanding invoices, the principal investigator knows that supplemental funds were not available to cover additional costs, and that there was no change in the scope of work to justify additional payments. The first step is to prepare a plan and present it to your management team. Next, approach the key personnel at your agency and companies that are depending on the final report. If possible, convene a meeting of these stakeholders and present your plan for evaluation.

It is recommended that you pay the ransom to obtain the report. To make certain that the report and its findings are what you expect, place the additional money in an escrow account to be released when the final report is accepted. After the report is obtained and the escrow has been released, prepare a request for additional funds to be sent to the project's sponsors, including the state agency. Your management team must endorse your request. Send copies to the representatives of the state legislature for your region and to the state governor's office. In that letter, you should present the details of your transactions with the contractor to justify the need for additional funds. It is important that all of the details are presented. If, for example, the principal investigator claims that the reason for increased cost is due to a change in scope, then you must acknowledge that claim and demonstrate its lack of validity.

The politics of government-sponsored research is a major underlying issue in this case study. Government agencies occasionally direct major environmental investigations to government or university laboratories. These government research activities often produce excellent work, but the agency contracting for the work has less control over the process. When the principal investigator is a tenured professor or research associate, the contracted study is more likely to be controlled by the researcher than by the sponsors contracting the research. This is not wrong or unethical, as long as the contracting agency is fully aware of the situation from the beginning. Sometimes, as in this case study, the public research institution bids against both public and private research institutions based on price, schedule, and technical merit. The public research institute competed for the project, and therefore it should be held to the same requirements as those of a private company. Using the advantage of tenure to get away with project mismanagement and having the effrontery to demand more money before releasing the final report is clearly unethical.

5.5 Expert Witness Hired to Create Reasonable Doubt

5.5.1 Dilemma

You are an environmental consultant for Down-Under Consultants (DUC). You have 25 years of experience in measuring, modeling, and controlling emissions to bring your clients into compliance with the local environmental regulations. As the agencies in your region seek to bring air quality into acceptable levels, they are proposing control strategies that involve allowable emissions that would require expensive control measures. Industry has found that these measures are excessive, expensive to implement, and an obstacle to current manufacturing practice. Occasionally this industry-agency disagreement is taken to court where experts on both sides confront each other.

In this case, industry employed three "leading" air pollution consultants to present various theories as to how and from where the pollution in question may have originated. In discussions afterward, one of the expert witnesses explained that these consultants were paid to suggest plausible—though one-in-a-million—possible causes for the pollution, and that there was nothing wrong with this. This response shows that the consultants lack the desire to ensure that justice prevails and have little respect for the environmental profession. In the end, owing to inconsistencies among the evidence presented, the judge disregarded all of the expert witness testimony and relied only on the accounts of actual eyewitnesses to make his decision. But how should environmental professionals behave in cases such as this? Is it unethical for an environmental professional to give exaggerated testimony to provide a basis for reasonable doubt?

5.5.2 Discussion

Our current issue concerns the ethical behavior of environmental professionals who are invited to appear as expert witnesses in court on environmental pollution cases. In these cases, the defendant, industry, or agency hires expert witnesses to give testimony and introduce the notion of reasonable doubt into the possible cause of the pollution. This dilemma considers whether it is unethical for an environmental professional to give exaggerated testimony in pollution prosecution cases to provide a basis for reasonable doubt.

The issue, in this case, is whether the testimony is honest, not just truthful. The first of Josephson's six pillars of character is trustworthiness. Trustworthiness encompasses the virtues of honesty, truthfulness, sincerity, integrity, fidelity, and promise keeping. Honesty means telling the truth, the whole truth, and nothing but the truth (as is demanded by the oath taken in court). An environmental professional acting as an expert witness in court to conjure up an idea that creates reasonable doubt, but is highly unlikely, is unprofessional, unless he presents the degree of unlikelihood associated with that idea. Similarly, giving dishonest testimony under the encouragement or direction of a defense attorney for the purpose of creating reasonable doubt severely damages the dignity of the environmental profession. This is especially true if the expert witness deems that the judge and jury do not have the knowledge or resources necessary to recognize the lack of validity of the expert's argument.

Another of the six pillars of character is responsibility. Responsibility comprises the virtues of accountability, leading by example, safeguarding integrity, and maintaining a good reputation. An expert witness has a duty to ensure that the court fully understands the testimony given and to not mislead the court in any way. This is to say that when there is reasonable doubt in the mind of the expert witness, that doubt should be expressed. The fundamental difference between the ethics of the environmental professional and the ethics of the legal professional lies in how *ethics* is defined. In this book *ethics* is defined as the difference between "what you have the right to do" and "the right thing to do." On one hand, legal behavior is doing what society says you have the right to do (e.g., evicting a tenant who is ill and unable to pay the rent). On the other hand, ethical behavior is doing the right thing (e.g., helping the tenant to find social relief to remain in the building).

5.5.3 Recommended Action

An environmental professional is responsible for being totally honest; this is true both inside and outside the courtroom. While an environmental professional's top priority is to protect the public, which includes developing and complying with regulations, this must be done with honesty and full disclosure in order to maintain the dignity of the environmental profession.

However, even experts can and do disagree because they use different approaches to estimate exposure. One may use direct measurement, while another may use modeling to estimate exposure. Then there is the current problem of global warming. It is difficult to conclude whether global warming is anthropogenic or not. Scientists and politicians continually debate the issue. In a courtroom and after plaintiff and defendant support testimony, it would be difficult for a juror to decide on what is the truth. The interesting point is that both sides believe that they are giving an honest portrayal of the phenomenon.

So to answer the original question of whether it is ethical to present exaggerated testimony to refute the theory that is the basis for regulatory action, the answer is a definite maybe if the argument is an honest presentation of the science and technology.

5.6 Colleague Seeks Help to Stop Supervisor's Suggestive Remarks

5.6.1 Dilemma

This dilemma is not confined to the environmental profession, but it is one that can be and has been encountered by professionals working in the environmental industry. The solution recommended below has been consulted with human resources professionals.

You are one of five engineers at Tests-R-Us (TRU), a small source testing company. TRU requires that its engineers and technicians spend 50 percent of their time in the field measuring emissions from industrial sources. The test department's working supervisor, Howard, to whom the engineers and technicians report, is a 12-year veteran of the testing business and has been with the company since it was founded. Howard has a strong ego and an "earthy" manner. You feel that his constant stream of dirty jokes and sexual innuendoes is inappropriate in the workplace, and it often makes you feel uncomfortable. However, since he has not directed any inferences toward you personally, you have avoided mentioning your discomfort to TRU's management.

You are aware of his comments and actions toward Shirley, one of the female technicians in the crew. He continually makes suggestive comments about her physical appearance and he purposely schedules work assignments so that he is with her alone at remote workstations. One day, Shirley comes to you in confidence and is obviously distraught. She just completed a test in a remote location where Howard was the sole test engineer. She says that Howard made her life miserable, but she has been hesitant to report him to TRU's management because she desperately needs the job. She asks for your help. How do you proceed?

5.6.2 Discussion

Clearly, the simple solution would be to have Howard stop this boorish behavior immediately and apologize to Shirley. The problem becomes how to persuade Howard to do this without risking your job or Shirley's.

While instinctively you may want to help a colleague in this situation, and even feel an ethical obligation to do so, reacting in a cavalier manner could have serious repercussions. You might think that a friendly, confidential conversation with Howard could resolve the problem without causing a major fuss. However, if you approach Howard with accusations of sexual harassment, you risk not only incurring his wrath and potential legal action, but if your involvement angers TRU's owner, also being fired. In a perfect world, Shirley should be able to confront Howard, her supervisor, and let him know that his suggestive remarks and inappropriate behavior make her very uncomfortable. Often in these kinds of situations, those responsible for the harassment do not appreciate how others receive their behavior.

5.6.3 Recommended Action

What might have begun as good-natured teasing has simply gone too far. Howard could theoretically apologize to Shirley and change his behavior in the future. But this is the real world. Since Shirley does not feel comfortable approaching Howard directly—and her pleas to you for help clearly indicate that she does not—you should suggest that she go in confidence to the TRU's management. If the company is large enough to employ a human resources representative, she should seek help there first. If not, she should be able to approach the owner or top manager without fear of losing her job. Advise her to document specific instances of the harassment and tell her that you would be willing to speak with management regarding your observations of Howard's behavior. You might also offer to discuss this issue with other colleagues in the group, but only if she is comfortable with that action (you should warn her that others in the group may not want to become involved). If she has the support of her colleagues, then the impact should be stronger. The owner should realize the seriousness of the situation and the potential liability that the company faces. The owner should also appreciate Shirley's actions in trying to defuse an unfortunate situation without seeking outside legal assistance. If the owner is not supportive and the company does not take appropriate action to resolve the situation, Shirley would be within her rights to consult with an attorney. At this point, you have no further advice to offer and Shirley must turn to the resources available to her to obtain appropriate outside counsel.

In summary, sexual harassment can have serious consequences for any company, large or small, and must be resolved in the appropriate manner and by qualified personnel. When Shirley approached you for help, your response should have been to:

1. Listen to her problem, be supportive, and offer to be a witness to relevant incidents that you have observed.
2. Avoid the temptation to intercede by talking directly to Howard.
3. Explain to her that this matter must be resolved in the appropriate manner and by the appropriate personnel at TRU, and that it is not something with which you are qualified to handle.
4. Give her your moral support and offer to speak with others in the group. If Shirley agrees, explore your colleagues' willingness to help validate her claim.
5. Suggest that Shirley report this matter to TRU's management. (If the firm is larger than the one described in this dilemma, suggest she report it to the company's human resources representative.)

6. Advise her that when she meets with TRU's owner, she should explain that she is interested only in putting an end to Howard's inappropriate behavior and resolving the issue without any outside involvement.

The owner should understand that Shirley's initial reaction to Howard's suggestive behavior may have been less than emphatic because she was afraid of offending her immediate supervisor and of losing her job. Unfortunately, Howard may have taken her "quiet" response as being receptive to his behavior. The point is that no one wants this kind of situation to escalate out of control, and therefore sexual harassment of any sort cannot be tolerated. Hopefully, prompt action can bring a satisfactory resolution.

5.7 Plant Manager Demonstrates Compliance with Incorrect Agency Lab Data

5.7.1 Dilemma

This case study considers a situation in which a regulatory agency's laboratory test results are more favorable than the permitting plant's own data. You are the environmental manager of ZZ Manufacturing, which has a National Pollutant Discharge Elimination System (NPDES) permit to discharge wastewater to the local sanitation plant. Per the requirements of the NPDES permit, the local water pollution control district (WPCD) staff collects a 24-hour composite sample of the discharge. The sample is split three ways, and each vial is closed with the WPCD seal. The WPCD keeps one sample for analysis in its laboratory, and the other two samples are left with ZZ. One of the two samples is analyzed at ZZ, and the other is sent to a certified outside laboratory. Proper chain-of-custody procedures are followed throughout the process.

The results from ZZ and the outside lab agree, while the WPCD laboratory's results are lower by a factor of 10. The outside lab reconfirms its results. According to the district's lab results, the discharge was in compliance with regulations; however, according to the other two lab results, the discharge exceeded the allowable level of pollutants. How do you proceed?

5.7.2 Discussion

Fear of fines and public reaction can lead employees of companies to cover up minor nonconformances with regulations. Technicians do not want to cause trouble or look bad to the boss. Chances are that the agency will not discover the violation, so why not? However, there is always the chance that the agency will discover the discrepancy at a later date. By confessing to, and correcting, small violations such as the one described in this dilemma, a company can build a sincere relationship that will pay dividends when handling more serious discrepancies that might occur in the future. In addition, as environmental professionals we want to do the right thing, not the thing we can get away with.

It is extremely important that ZZ's management drum this philosophy into its staff. This dilemma is not earth shattering, but it is the kind of situation that comes up frequently and needs to be handled properly and consistently, so that the more consequential dilemmas are also handled in the right way. Cheating has been compared to the actions of a boa constrictor. You get away with one small action and then another until a major issue arises. You are tempted to say, "Why not?" A boa constrictor winds itself around its victim. Each time the victim exhales the snake tightens its grip until the victim can no longer breathe.

The right thing to do is to make the discrepancy among the lab test results known to the WPCD staff. This demonstrates your trustworthiness, responsibility, and fairness; in the long run, it will gain the district's respect, which can only benefit your company in the future. The district may have a calibration bias in its lab, and should be made aware of it so it can perform its duty of protecting public health by properly enforcing regulations. The WPCD management will appreciate your honesty, and the next time there is a discrepancy in the results, it will work more cooperatively with your company. For example, the next time its results are high and your results are low and confirmed by the outside laboratory, the district is more likely to consider your results for compliance.

5.7.3 Recommended Action

Before contacting the WPCD, however, it is essential that you determine what caused the high discharge concentration. Assuming that the process was at fault, the cause must be determined and corrected, or a means for correction determined and initiated. If ZZ goes to the WPCD with a corrective action in place, or at least a plan of action, it is highly unlikely that the WPCD will give ZZ a notice of violation for the discharge. To notify the district of the discrepancy, the highest official at ZZ should contact WPCD's executive officer. Ideally, the agency's executive officer will appreciate ZZ's honesty in self-declaring this situation and in taking immediate steps to correct it.

While many processes are required to have continuous emission monitors that record any out-of-compliance situations, it can build a relationship with the district's inspectors to call these events to the attention of the inspector, even if an occasional exceedance is acceptable. This demonstration of honesty has very little risk and will keep the inspector reporting in your favor.

5.8 New EH&S Director Has to Bring Plant under Compliance and Keep It Open

5.8.1 Dilemma

You are the environment, health, and safety (EH&S) director with the global high-tech firm Microchips-R-Us (MRU). With increasing competition from overseas and rising costs at home, MRU has begun relocating some of its plants outside the United States in an effort to reduce costs and increase profits. One of the company's divisions, located in the Southeast, has shown less than desirable performance in recent years. As a result, MRU has assigned a team of key company executives to assume the management of that division and its three plants. MRU is the major employer in the region, with 5,000 workers at the three plants. The new management team has been briefed to improve the division's operations; otherwise, three plants will be transferred to new locations in Asia.

You have been selected to be the division's new EH&S director, and have relocated your family to be near division headquarters. You meet the resident EH&S staff, who average 15 years with the company, and present them with your audit plan. After three weeks of surveying plant operations, checking records, and interviewing plant personnel, you draw up a list of noncompliance items. The list is long and will require major capital expenditures to bring the three plants into regulatory compliance. Key deficiencies include several serious toxic release issues affecting both worker and public safety. You learn from your EH&S staff that they are well aware of the problems. They inform you that their memos concerning noncompliance were ignored by the former management, and that any complaints from employees regarding health and safety were deferred with the excuse that there was no money available to fix the problems. Besides, the former plant manager held prestigious voluntary positions in local government and always arranged to personally meet any agency inspector who might show up at one of the plants.

It is a well-known secret that the division is in trouble and that any significant expenditure could mark its death knell. You meet with the new management team and learn that these plants have some unique proprietary processes and personnel experienced in their operation. It would not be easy to transfer those operations to another location, let alone a non-English-speaking country. The management team would like to keep these plants open, but MRU's CEO will not accept excuses. He has promised the company's board of trustees that the new management team will fix this problem. How do you deal with the situation?

5.8.2 Discussion

The dilemma here is clear. On one hand, employee and public health and safety are at risk, while on the other hand, the means necessary to resolve this problem could seriously affect the lives of some 50,000 people. Not only would the plants' closure affect the local communities, but also your own family and career. You have moved to the area specifically for this job. If the plant closes, you will have to relocate, and perhaps look for a new job. It is evident that these plants have redeeming qualities in their proprietary processes. Clearly, a team effort is needed to make the plants successful and more cost-effective in order to afford the necessary EH&S upgrades. There is also, of course, the major issue of dealing with the regulatory agencies. Records have not been properly maintained, reports have been falsified, control equipment has not been updated, and permits have not been obtained. There is a great likelihood of serious penalty, including hefty fines or even imprisonment.

5.8.3 Recommended Action

Your first step as EH&S director is to work out a solution to the problem with the help of the division's new management team. You must develop a close working relationship with the division manager. Your efforts should also include operations, maintenance, and quality control staff, as well as corporate marketing staff. After an extensive site investigation, a clear and precise recovery plan should be prepared and submitted to the CEO for acceptance. Ideally, the plan should demonstrate that with moderate investment (e.g., loans or corporate funds) the division could turn itself around and become profitable, even after paying off the equipment upgrade costs and regulatory fines. If this result cannot be achieved, then the future of the plant is dim and a phase-out plan will be needed to mitigate the impact on the community.

One of your responsibilities will be to deal with the regulatory agencies. Because of the seriousness of the issues, it would be advisable to also seek legal counsel. You should engage the help of an environmental attorney who has close working relationships with federal, regional, and local agency officials. The attorney will likely require that you make a full disclosure of the plants' violations and MRU's plans to comply with the regulations, and that you present an explanation for the violations and a full assessment of the public hardship that might occur if the plants should close. By self-reporting the noncompliance issues, the consequences should be mitigated. Keep the initial negotiations with the regulatory agencies as private as possible. The last thing you need is media coverage. Although the news will come out eventually and must be handled accordingly, including the corporate marketing staff in your plans will help prepare for this eventuality.

Because of the potential adverse impacts on the local community, political action may be required. Your management team will need to work especially hard on public relations in the region. You may find yourself in the role of the head "cheerleader." There are often employees who are active in local community affairs. Recruit them in the process, and allow them to help gain public support. Be aware that there may be self-serving citizen or political groups that will try to capitalize on your company's problems. It is desirable to gain the support of the local community before such groups gain a foothold.

It is not uncommon for companies to find that competition at home or overseas is driving them out of business. As the competition intensifies, market prices for products decline, and plant managers begin cutting corners to save costs. Because this is a gradual process, the company may find itself in a mess before a full recovery plan can be prepared and executed. In this dilemma, one area that was neglected was compliance with health and safety laws. This is never acceptable. Environmental professionals must take every step necessary to protect public health and maintain strict compliance with environmental regulations at all times, or their firm can suffer severe consequences.

5.9 Agency Employee Moonlighting as Private Consultant

5.9.1 Dilemma

You are the deputy air pollution officer in charge of a branch office of a (U.S.) state air agency with 50 environmental professionals and an equal number of support staff. Your office's duties include assisting the headquarters office in the development of regulations, issuing air permits, enforcing air pollution control regulations as reflected in those permits, and monitoring regional air quality. Your staff includes environmental scientists and engineers, computer modelers, historians, anthropologists, and writers.

While reviewing a permit application, you notice that the supporting health risk assessment was performed by Tom Jones, a member of your staff. You speak to Mr. Jones, who admits to working part-time for UTC, the consulting firm that prepared the permit application. Tom explains that he occasionally runs models when the UTC's regular modeler is busy or on vacation by inputting the emissions data provided by the UTC's engineer and formatting the output to conform to agency guidelines. Most of the time, Tom says, the work for UTC involves projects in other states. He also explains that he does this work at home and on his personal time and feels that there is no conflict of interest. Tom also mentions that other staff members at the agency are moonlighting in the same way, adding that they never work on projects with which they are involved at the agency. How should you deal with this issue?

5.9.2 Discussion

You determine that your agency has no specific ban on this type of employee moonlighting work, as long as it does not conflict with the agency's mission. Assuming that you investigate further and find no evidence that Tom (or any other staff member) has exerted any inappropriate influence on the agency's permitting or enforcement processes, is there indeed a conflict of interest or ethical problem here?

The trustworthiness of this particular agency staff member, Tom, becomes suspect when he is found to be employed by a firm applying for a permit from your agency. He claims he will not accept work if it involves a project in the agency's jurisdiction. However, even if he performs the modeling work totally objectively, uninfluenced by UTC or its clients, there is still the appearance of a conflict of interest when the consulting firm undertakes any advocacy action with the agency on behalf of this or another client.

There are both proponents and opponents to most permitting applications. Consider the situation in which a permit application submitted to your agency involves a contested risk assessment. Imagine the embarrassment if Tom is the agency's expert and the opponents to the permit learn of his involvement with UTC, which is submitting the application, even if he was not directly involved in the case at hand?

5.9.3 Recommended Action

As the deputy air pollution control officer (APCO) in charge of this branch office, you should assign an assistant to interview each of the environmental professionals on your staff to learn the extent of their moonlighting activities. Learn the names of all the consultants who have used agency staff members' services on a private basis, and determine if there are any projects involving your agency in which they have participated. This information will prepare you for any adverse actions that might arise in the future.

Next, contact the APCO at the state agency. Explain the situation and request a meeting with the agency's senior staff. It is likely that this situation exists in other branch offices. It is apparent that a significant policy action must be considered. The key issues to be resolved are (1) preventing the future appearance of conflicts of interest, and (2) dealing with the wake of legal actions or public relations issues that might arise as the result of prior staff moonlighting.

A formal restriction of agency staff members working for consultants with any advocacy action with your agency should be promulgated. Consulting firms that have used agency staff services in the past should be advised of the impropriety of this relationship and of the agency's new policies. Each staff member who moonlighted in these cases should prepare an account of the services he or she performed and also sign a statement asserting that he or she feels no obligation to favor the consulting firm who used his or her services. Finally, if possible, ask an executive at each consulting firm involved to sign a statement recognizing the agency employee's disclaimer.

The most important thing to remember in handling this type of situation is to inform and involve senior management immediately and deal with it as an executive group to achieve the right resolution with the least impact on the stakeholders. While your employees have a legal right to moonlight, this is not the right thing to do.

5.10 Firm Sponsors Clients on Exotic Speaking Engagements

5.10.1 Dilemma

As the executive vice president (EVP) of a large international environmental consulting firm, Compliance Are Us (CAU), you notice that your highest performing division has an exceptionally large marketing travel budget. Despite incurring this large expense, the profitability of the division by far outperforms the rest of the firm. You look further into this operation in an effort to export some key performance techniques to your lower-performing divisions. You quickly learn that this division has been sponsoring executives from two of its major clients to go on speaking engagements. For example, you learn of one trip to Manila to speak to an engineering society about creating wealth, another trip to London to speak at a world conference on finance, and similar travel to meetings or conferences in Honolulu, Rio de Janeiro, and New York. Each trip lasted a week or more, and all expenses were paid by CAU. You also learn that the clients take their spouses on these trips and all expenses are paid by CAU.

You speak to the division vice president about this matter, who explains that these clients account for 60 percent of his division's consulting activity, and that CAU has an exclusive arrangement with the firms to handle all of their environmental consulting. He says, "We have performed very well for them and in return they accept our billing rates without a hassle. Our relationship extends back nearly eight years, and I have made it my personal duty to keep in close contact with their chief executives. They are wonderful people. The men are great speakers, and over the years, we have provided opportunities for them to speak around the world. They are our best supporters and often refer us to new clients. I think they are well worth our investment." That investment has ranged from $22,000 to $58,000 per year, increasing each year along with the division's profitability.

As corporate EVP, you ask the division vice president if he is comfortable with the fairness of gaining this most favored client status by offering free trips around the world. He says, "Sure, it isn't just the trips that keep us on board. We give them priority service and use our best project staff on their work. I personally get involved when they need my connections at the various agencies to expedite permits through the system. It's not like we're doing anything illegal. These trips are not kept a secret, and I believe that their boards of trustees are aware of our sponsorship, although that fact is not advertised." From subsequent interviews with the division's senior staff, you come to believe that this activity is generally known by CAU employees, but not openly discussed. What action, if any, should you take?

5.10.2 Discussion

This dilemma illustrates how environmental professionals must be aware of their core ethical values of trustworthiness, responsibility, fairness, and citizenship in their pursuit of business growth and marketing. When do seemingly harmless actions, such as borrowing pens from the office supply cupboard or fudging expense accounts, become stealing? Similarly, when do free business lunches and drinks become inducements? Certainly, clients enjoy doing business with people with whom they have a friendly relationship. It is not unethical marketing practice to gain the friendship of a client. Similarly, paying for lunch to better acquaint a client with your firm is not necessarily unethical. This dilemma involves high-level executive clients, where a lunch meeting is valuable, considering the price of the executive's time. Indeed, picking up the tab on a business lunch is even permitted by the U.S. government, which has established a standard price limit of $25 per lunch guest—there are even some restaurants in Washington, D.C., that offer special "Government Guest" lunches on the menu priced at exactly $25. Then again, an executive with strong ethical convictions would elect to pay for his or her share of the bill to avoid even the slightest presumption of impropriety.

At first glance this situation may seem like a valid means of doing business. But further consideration shows that client executives are receiving travel perks and their companies are paying for them indirectly. The rates that your company charges these "major clients" are inclusive of costs of these trips. An analogous situation would be a purchasing agent handling a routine purchase of production hardware. What if a vendor company offers a discount on certain of its products provided that a minimum quantity is purchased? To keep the good side of the buyer, the salesman offers to charge his company the full price that it usually pays for those products and pass on the amount of the discount to the buyer. Is that a bribe? This case is more clearly an unethical arrangement. If the division vice president were to give a check to each of the executives to cover the travel and make the speaking arrangement, would you then consider the checks a bribe?

It is one thing for the company to offer a discount on a product to encourage a buyer to make a purchase. Bargaining on the purchase of an expensive item like an automobile is a normal activity. But, if your agent obtained a discount and you paid the full price, you would feel cheated when you learned of the deal. Now, if the two traveling executives were actually the owners of their respective companies, the issue of a bribe would no longer be valid. But as a company representative, "an executive" is part of the company entity; even if they are owners, using other people's money, the company is paying for the trip, not you. Companies have been known to invite clients to a day on their fishing boat and they charge the expenses to their marketing budget. Whether or not this is a bribe would be the subject of an ethics workshop discussion.

5.10.3 Recommended Action

This is a difficult dilemma to resolve and may require the help of a professional ethicist, who can present an objective view of the situation. As CAU's EVP, your first step is to obtain the support of the firm's president. Next, schedule an all-day, off-site meeting of the company's officers and senior managers. Do not disclose the purpose of the meeting in advance, but begin the meeting by presenting the details of your investigation. Be careful not to attack any individuals and avoid passing any judgment, but instead present this as a company culture problem that you wish to address without delay. Then, introduce the professional ethicist who will facilitate the meeting. Clearly, the objective of this session is to devise a means to end the practice of providing large-scale gratuities to your major clients, with no or minimal negative impact on the firm. The means to accomplish this is what should emerge from the group dynamics of the all-day meeting. One approach is to take measures to ensure that the work performed by your staff is reviewed sufficiently to guarantee that the best work was done on each assignment. Have your senior management member that has close relations with the client tactfully advise the client's management of this procedure. Be alert for any dissatisfaction or negative reaction and address it immediately. This is clearly a difficult assignment and will require all of the company's managerial support to be accomplished successfully.

5.11 Young Professional Learns the Downside of Not Blowing the Whistle

5.11.1 Dilemma

You are a recent environmental science graduate of Mid-West College who has been hired by ACME Power, a large electrical utility company. You are stationed in an eastern New Mexico power plant with 200 employees. The plant is 30 years old and many employees have been there for 20 or more years. The plant has the latest in emissions controls, including continuous recording instrumentation. Every colleague you meet is friendly, and your new boss, the environmental manager, assigns you to record keeping. You spend several weeks with a colleague named Pat, who you are to replace and who is being promoted to environmental manager at another plant. During that time, you spend each shift getting acquainted with your record keeping duties, as well as with the personnel in various activities in the plant. Meanwhile, Pat is busy completing reports, but takes the time to show you where to obtain the data. The local air agency inspector drops by to say goodbye to Pat. You observe that Pat has a warm relationship with the inspector and the plant has an excellent record of compliance.

On the day Pat leaves, he stops in your office with these parting thoughts: "Listen, we need to cover a few more items. The 'old man' (plant manager) is a tough old bird who wants to avoid any regulatory problems. He keeps the output at the demand level and has safety as his primary concern. He wants no problems with the agency and expects your reports to be consistently within the allowable. We have never had a notice of violation, but confidentially, we do occasionally have to fudge the emissions traces, especially during load swings. The instrument technician can show you how we handle the adjustments. Oh, and the fuel sulfur levels need recertification to keep them in spec. The lab does that pretty much automatically. I always check the reports to be sure that there is consistency between the fuel sulfur and the sulfur dioxide emissions. You'll get the hang of it soon."

Four months later, you hire a test engineering firm to conduct an annual audit. Its report shows that the plant is out of compliance and a copy of the report has been sent to the local agency. You are surprised and take the findings to the lab manager. You learn that your company's regular test engineering firm was hired, but they sent a new crew that had not been to the plant before. The lab manager did not notice the personnel change because Pat had always handled the coordination. The new crew was not acquainted with the adjustments needed to obtain acceptable test results. On top of everything, you receive an email from the agency indicating that it will be investigating the malfunction.

You make a call to Pat at his new location to ask for advice. His initial reaction is, "Look, I told you that you had to screen the data before submitting it to the agency. Have you told the old man? You better work with the instrument technicians to get a story together. The plant has needed a new nitrogen oxides catalyst for two years, but our maintenance budget didn't schedule it until next year. I suggest you get to the old man first. I'll help you as much as I can."

5.11.2 Discussion

It is apparent that Pat and the plant staff were never sensitized to their first priority of protecting public health. The plant manager has worked his way up from plant operator by being a strong leader. He is likely to be a whatever-it-takes type of person. He is good to his family and will not overtly lie, cheat, or steal. But dealing with the government is an area where he is willing to play fast and loose. Pat was a team player, and his career benefited from his willingness to keep the plant manager happy. The problem is that fudging data is not only unethical and illegal, but also will eventually be discovered. If the cover-up is not inadvertently discovered, as it was in this dilemma, an employee will eventually leave the plant for another job and may mention how things were handled at his old job to new colleagues.

5.11.3 Recommended Action

You must set the record straight, correct the functional deficiencies, and minimize the impact on all stakeholders. The specific actions will depend on how the plant manager reacts to the situation. You need to gain his confidence in that you are trying to support him to do the right thing. The first thing you should do is determine the regulations involved and the possible regulatory actions that might be taken. Then, you need to gather the records that have been falsified, as well as the original unchanged versions. This information will be needed regardless of what action is taken. The plant manager should immediately contact the company's attorney, who will no doubt take the lead by collecting all of the relevant data and researching the regulations. It will be important to capitalize on the good relationship the plant has with the agency.

It will be essential that you are completely honest in relating all of the facts and situations with which you were involved. Honesty involves not only telling the truth, but the whole truth. Do not be persuaded to tell a story that is anything but truthful. This is a difficult situation that you could have mitigated by blowing the whistle when you first learned of the data altering. You should have gone immediately to the plant manager and made him aware of the situation. It is uncertain how you would have been received. Now, you need to utilize all of your tact and personality. Show him the regulations and potential fines he is facing. Suggest that he speak to the company attorney or with outside counsel. You may risk losing your job, but that is part of the meaning of taking responsibility.

5.12 Supervisor Uses Company Resources for Book Not under Company Auspices

5.12.1 Dilemma

You are an environmental professional working as a mid-level nonsupervisory member of a 25-person environmental compliance department of Huge Paper Company (HPC). You are three years out of school with a bachelor's degree in civil environmental engineering. Your experience in the department has included a range of assignments involving emissions control systems, hazardous waste remediation, permitting, and modeling. You feel that you are well respected and your work has received compliments from the project leaders and Harry, the section head to which you officially report. Harry has not only complimented you, but has told you that you will be given a leadership role on a project in the near future. You are satisfied with your pay raises, and your aspiration is to advance through the department.

One day, you are given an assignment by Harry to collect data from the department files on emissions of certain air pollutants and the related control devices that were used on the stacks involved with those emissions. When it is time to complete your time card, you go to Harry and ask for the appropriate project number to which to charge your time. You are told to simply charge your time to the project you worked on last week. You are puzzled because last week's assignment had no relationship to emissions. After work, you go out with some colleagues for a drink. You happen to mention your confusion about the time charge number to another member of your department. He says that Harry is writing a book and has a contract with a publisher for a due date next year. He uses his leadership role in the department to obtain free research support from his subordinates. You learn that Harry spends his evenings and weekends working in the office on his book. You also realize that when you have completed your assignments, you review your findings with him on company time, sometimes for hours. Finally, you find the courage to ask him if the department is supporting his writing of the book. He brushes off your inquiry, leaving you in a quandary. How do you deal with this situation?

5.12.2 Discussion

As a young person in the department, you are naturally interested in making friends and avoiding personal conflicts, especially with your supervisor. You wonder how much freedom your colleagues have to use the department's resources to do work not directly related to company business. Your instinct tells you that unless Harry's book will be published under the company's auspices, his preparation for it should be done outside of working hours. And if the company's data are to be published, management must grant permission. As a section head, Harry may think he is high enough in management to make that determination, but this is very unlikely.

Do you feel a responsibility to report him to HPC's upper management? But what if he has the company's permission? Your challenge would make you look like a chump and certainly risk losing Harry's friendship. Is this akin to his asking you to steal information? What if he asked you to take a computer to his home without telling anyone? Would you ignore either situation to stay on Harry's good side, or do you report him? Before reading on, try to answer these questions for yourself.

A colleague discovered behaving unethically in the workplace is a sensitive situation. You may or may not have a personal relationship with this person, and the person may or may not recognize the action as a breach of ethics. It may even be that you have been mentored by this person, and therefore are readily convinced that the suspect behavior is perfectly acceptable. Nevertheless, dealing with this type of situation can prove to be a real test of your integrity, responsibility, and diplomacy.

5.12.3 Recommended Action

First, because of its sensitivity, you need to gather as much information as possible before making any accusations. You need to have a conversation with someone in the organization that you can trust to discuss the situation confidentially. Find out how the book preparation is being financed. Is it a well-known secret that everyone ignores? Are other members of the staff "helping" Harry in the same way? If you get the impression that the staff realizes that what is going on is wrong, then speak to others to assess their general attitude. Your bringing it to their attention should bolster your support in going forward. If you come to the understanding that management condones Harry's actions, then you can back off. It would be good to speak to Harry to better understand the situation. Does the company derive a secondhand benefit from Harry's authorship? Are the data he uses public information? When seeking this information, avoid being accusatory. Explain that you are interested in his work and want to have a better understanding of the final product so you can help him more effectively in the future. If Harry's work in this regard is on the up-and-up, then you can relax.

If, on the other hand, you determine that the consensus is that Harry's book is being done without the knowledge of HPC's upper management, then you have a responsibility to take action. To minimize the possible impact on you, you should try to gain the support of your colleagues. Explain to them that it would be better to stop Harry's taking advantage of his position before management discovers it. Also, explain that each of them will gain the trust of management if they handle the situation properly, and that by working as a group you all reduce the risk of any potential retaliation from Harry.

Once you have gathered together a team, you should approach Harry with the objective of making him understand the problem. Be as friendly as possible, and explain that his staff wants to avoid any problems from management if it learns about the book's preparation. Perhaps you could suggest that he try to sell management on the notion that his book will have a net benefit to the company. In any event, the "free" work must stop. If all else fails, you must take the allies you can muster and go to upper management or the human resources department. A good human resources staff member should know how to handle this situation with the least impact on the stakeholders.

5.13 Production Manager Installs New Furnace in a Hurry without Proper Permit

5.13.1 Dilemma

You are the environmental supervisor for the Widget Manufacturing Company (WMC) plant in the Midwest. For the past two years, business has been in a slump and your WMC's output has been reduced significantly. Your environmental staff has been eliminated, and despite your 15 years of service with the company, your title of manager has been downgraded to supervisor because you are now a one-person group. You are pushed beyond normal working hours to maintain routine compliance actions. Because of your intense work schedule, you find that you cannot make routine patrols of the plant as was done by your staff in the past. One day, while walking through a part of the plant you seldom frequent, you spot a new machine installation. On further inspection, it becomes clear that the unit is an annealing furnace in full operation. There is no operating permit for that furnace.

You confront Jack, the production manager, who explains that the plant received a rush order for huge metal parts that required a larger annealing furnace. In the panic of getting the project moving, he found a used furnace that he could install immediately and failed to go through the normal process that would have ensured that you be notified about the change. Jack suggests that since the older furnace has hardly been used over the past two years, you merely hang the existing permit on the newly installed furnace. "I'll have my guys put the new furnace in the old one's place and put the old one in off-site storage or up for sale. We could even change the nameplate. The air pollution agency inspector, who doesn't really come around anyway, will never notice the difference." In spite of the pressure from the production manager, you do not want to accept this approach. You suggest that Jack and you go to the plant manager.

5.13.2 Discussion

You know Jack. He is your colleague and is basically a good character. In a tough business situation, it was understandable that he was obsessed with producing a quality product and meeting the customer's schedule. Now that the problem has been discovered, however, the objective is to do the right thing with the least impact on all stakeholders.

The answer may seem obvious. You know what the law requires, but is following the law to the letter the right thing to do in this case? The pressure on you is significant. The plant is struggling and the last thing it needs is to be charged with a violation and have to pay a significant fine and possibly suffer a worse penalty. The plant is operating on a marginal basis and all of the employees are stakeholders at risk if the plant shuts down. Your family would suffer if you lost your job. If you take a rigid compliance attitude with management, they might decide to replace you with someone who is more flexible.

Following Jack's plan would not only put the company at risk of a breach of agency regulations, but possibly a more serious offense, maybe even a felony. Thinking that no one would ever find out is naïve. With the plant on a downsizing basis, there's bound to be some disgruntled ex-employee who would be willing to blow the whistle to the agency. If you go to the agency yourself with an after-the-fact permit application, you'll likely be issued a violation and be subject to the disciplinary measures issued by the hearing board. Before you go to the plant manager, you'd better have a plan.

5.13.3 Recommended Action

When you meet with the plant manager, you should suggest two approaches. The first approach would be to have the plant manager suggest that the company's attorney meet with the agency prosecutor to explain the situation. The attorney should use the firm's compliance record to its advantage. Hopefully, this record has been unblemished in the past and documentation supporting this performance can be provided. Then, the attorney should discuss the financial burden that this compliance has had on the company's profits. Clearly, the objective is to create a sympathetic mood in the prosecutor. At the appropriate time, the attorney should introduce the permit problem as a result of the plant trying to keep operations going with a reduced staff. It would be unreasonable to expect the prosecutor to merely dismiss the matter. But if you have provided your attorney with a permit application that can be submitted at this meeting, then in many jurisdictions, there is a possible routine disposition with a minor penalty, like doubling the permit fee. Also, in many jurisdictions, the permit application would serve as a temporary permit to operate.

The second approach would be to have the plant manager meet with the chair of the local air pollution agency board, provided they have established a friendly relationship. If this is not the case, then you should encourage the plant manager to establish this friendly working relationship so that he may handle similar issues in this way in the future. Suggest that the plant manager arrange an informal meeting with the chairperson. If these two executives can reach an agreement on the disposition of the annealing furnace permit, then you can process the permit application per that agreement.

Environmental professionals must do their best to comply with environmental regulations and protect public health. However, things do not always run smoothly, and they may be faced with recovery actions like the above scenario. This situation is one that is most likely to occur in a down business cycle when the plant staff is under pressure. Therefore, it is important to consider how inappropriate actions are sometimes contemplated in an attempt to find a shortcut around a regulatory problem. It takes courage in such circumstances to enlist the support of colleagues and supervisors in order to do the right thing.

5.14 Equipment Supplier Offers Incentive to Make an "Objective" Recommendation

5.14.1 Dilemma

As an American independent consultant with extensive experience in the environmental field, you are engaged by the Environmental Control Agency (ECA) in the country of Miland to advise on the technical aspects of a technique that is used in the United States and considered to be a maximum achievable control technology (MACT). Although the MACT technique has been in practice for more than 25 years in the United States, it remains a standard in the industry. However, the performance requirements (i.e., allowable emissions) differ in severity in different areas of the country, depending on the attainment status of the individual areas. This can make the emissions control process expensive, so the industries that use it are continually searching for alternative techniques that can lower operating costs, as well as further reduce emissions.

Miland has recently discovered a disturbing cancer rate in its larger industrial cities, and as a result, the country's ECA is investigating the use of emissions controls practiced in Europe and the United States. Miland's ECA has decided to pattern its control regulation after the New Jersey regulation, which is among the most severe in the United States. Due to your expertise in this area, you have been a consultant in the development and use of this process and have authored several publications in the literature. You are approached by ECA to advise its staff and Miland's regulated community on the details of this practice.

At the same time, you are aware of the emergence of a new technique for the control of emissions that may replace the current MACT standard. Several U.S. industries are in the process of evaluating the new technique, but cannot use it commercially until it has been approved by state and local agencies. You inform ECA of the new technique, but ECA is reluctant to implement the new practice until it has been approved by U.S. agencies.

You are convinced that the new process is viable and feel that it will soon be approved for use in the United States. You contact the various technology developers and suppliers to obtain performance data to support your intuition, which you then pass on to ECA. You also provide this information to Miland's regulated community. In the course of meeting with the U.S. technology developers and suppliers, one of them, Controls-Are-Us (CAU), appears to you to be the most promising. Through many exchanges with this firm, you become familiar with its staff. One day, CAU's marketing director visits you and mentions that his company is frustrated with the regulatory delays in the United States. He says that his company would like to promote its technology with ECA, since Miland is implementing a new program. Because of your professional connection to the Miland agency, he offers you a bonus incentive for any of CAU's equipment that is sold there in the next year. Is it ethical for you to accept CAU's offer?

5.14.2 Discussion

First, it is important to note that CAU is only one of several technology developers that have equipment with the potential to be used for the new control process in Miland. You consider CAU's equipment to be the most promising and have communicated as much to your contacts in Miland. However, you have also advised them that they should carefully evaluate each of the suppliers' equipment. Second, you have been hired to consult on the implementation of the existing practice in the United States. The issue at hand is whether you can exercise good faith in your consulting relationship with Miland if you accept CAU's offer. Furthermore, do you hold CAU's offer against the company? In other words, is this offer a bribe?

You owe your clients in Miland an objective evaluation of their options. They will ultimately make the final decision by evaluating the security of using proven technology against the risk associated with the significantly more cost-effective, but unproven, new technology. Your responsibility is to provide an objective appraisal of both approaches without demonstrating any preference. Even if your analysis indicates to you that the new technology is superior, you must not recommend one technology over the other, but merely point out the pros and cons of each. And even if you feel that the CAU technology is the most advantageous, you must be cautious that you do not "push" this technology.

As to the bribe aspect, yes, technically it can be considered a bribe. Whether or not you feel that you can be objective after you reject it, the offer introduces a dichotomy in your dealings with CAU. You had a preliminary assessment that this technology would be the most advantageous, and you had good feelings toward the company. Now, CAU has created an ethical dilemma for you. If you accept the offer, then it is tantamount to accepting a bribe. If you decline the offer and punish CAU by not supporting its technology, and if indeed it is the best one, then you have done a disservice to your clients in Miland.

You are aware that U.S. suppliers have been successful in securing contracts with foreign companies by using bribes. This practice is illegal, but seldom discovered or prosecuted. In this case, it is unclear whether the CAU marketing manager is aware that his offer to you is a bribe. Similarly, the CAU top management may not be aware of the offer. So how should you handle the offer?

5.14.3 Recommended Action

You must decline the offer and defer your response to the CAU marketing manager by finding an excuse to interrupt the conversation. Take time to discuss the issue in confidence with a colleague, if necessary. The important thing is to recognize this offer as a possible ethical issue. In your next meeting with the CAU marketing manager, you should proceed as follows: "Regarding the offer you made the last time we met, I cannot accept it. I realize that you made it in good faith, but I would not be able to proceed with an objective evaluation for my clients in Miland if I were to accept your offer. Even if my clients make the ultimate purchase decision and they decide on the CAU technology, I could not accept a bonus from you. I hope you understand and we can drop the issue immediately. I want to believe we both learned something from this situation and can move forward with this as a nonissue."

5.15 Campaign Manager Advises You to Oppose Environmentally Sound Project to Win Election

5.15.1 Dilemma

You are a professional whose commitment to civic virtue and citizenship has led you to become a member of the city council of Centerville, a medium-size town in which you live with your family. You are now running for reelection. Your public position on environmental issues has always been insightful, middle of the road, and noncontroversial. A local power plant in your city that is owned by a national energy company proposes to use hazardous waste as a supplemental fuel. The local plant has an excellent environmental and energy conservation record and, in your opinion, is sparing no expense to build a top-notch facility with a complete emissions control process and a continuous emissions monitoring system. The plant has been permitted and has met all of the regulatory requirements. The power plant will transport the hazardous waste fuel from its various sources using specially enclosed transporter vans to ensure that no waste will be dropped in any residential or commercial neighborhoods. You have studied the project plans and interviewed the plant's environmental consultants, who show you evidence of successful projects in other areas of the country. You are firmly convinced that the proposal is sound and that it should receive the city council's conditional use permit (CUP). Your political opponents have come out against the proposal (for the usual Not in My Back Yard [NIMBY] reasons, i.e., toxic waste transport on local streets, toxic air emissions, etc.), and this element is becoming a major campaign issue. Your campaign manager and the local party leaders advise you to oppose the project, for fear of losing the election. The council vote is one week away and the election is two weeks away. How do you proceed? Do you (1) ignore the issue, (2) come out against it to neutralize the opponent but give it little lip service or (3) support it, attempting to shows its benefits?

5.15.2 Discussion

In this dilemma you are an environmental professional with collateral position on the city council in the city where you live and are faced with a dilemma that would affect your reelection chances. With the help of your environmental professional colleagues you have been presented with unbiased evidence that power plants burning hazardous wastes, located throughout the country, are operating well and in compliance with air and waste regulatory requirements in those locations. The issue of traffic on the city's streets has been controlled by enclosed transport vehicles that operate only in nontraffic hours. But what of your campaign manager's and party leader's belief that the plant is a pivotal issue that could cause you to lose the reelection? You believe in the project and have learned the financial benefits to the city in terms of lowered electrical charges. You could come out against the plant and then change your position after being reelected. The issue is whether that would be the right thing to do when you already believe in the plant.

Coming out against the plant during your campaign would not be honest and would violate your responsibility to the citizens of the city, who will in the long run benefit from lowered electrical rates. You would lose the respect of the utility engineers who have spent so much of their time on the plant design and operation plan. And if, after the election, you reverse your position, you would lose the respect of the citizens, which could affect you in the next reelection.

Furthermore, political advisors are only interested in getting you elected. The public is seldom one issue oriented. If you can demonstrate in your campaign that you are interested in the welfare of your constituents by pointing out the advantages of this power plant, and that in your judgment as an environmental professional you can endorse it, they will ultimately support you.

5.15.3 Recommended Action

Do not be swayed by political advisors who have based their decisions only on polls of uninformed citizens that oppose the project because it seems too risky. The responsible thing to do is to spend the next two weeks speaking of the benefits of the project to the community and campaigning hard to get your message out. Contact the power plant management to support you with ads. Cite the successful waste-to-energy projects around the country.

Contact managers in those cities and get information regarding the issue of the truck traffic and how it is handled. Recruit the city's chief engineer and possibly local regulatory agency leaders to back up your beliefs. Learn the names of the leaders of the dissenting group of citizens and challenge them to a debate. Demonstrate that you care for the citizens of the city, and that this plant will give them employment opportunities in addition to being a benefit to the environment. Then you will feel comfortable that voting for the project is the right thing to do.

In discussing this dilemma, an assertion is often made that you can do more good for the community by being reelected than by campaigning on this one issue. You should concentrate on convincing your city council colleagues so that they can educate their constituents and let the industry advocates continue their campaign for project approval by having the public encourage their respective council members. At least that way, you may live to fight another day. Politics can be tricky!

5.16 Reporting a Toxic Solvent Leak under a Production Deadline

5.16.1 Dilemma

You are the environmental manager for one of ACME Machine, Inc.'s (AMI) manufacturing plants in Centerville, Idaho. The plant has just received a large, rush order that AMI's manufacturing manager, George, tells you will require this facility will be working nearly 24 hours a day for the next month to meet the delivery schedule. On one of your routine rounds of the plant you notice a small, continuous chlorinated solvent leak from a tank serving the hydraulic press that is the key machine operation for this project. You know that this fluid is a listed hazardous pollutant. You call this to George's attention and ask for his plan to stop the leak or collect and properly dispose of the leaking fluid. George's response is, "Fortunately, the leak is next to a floor drain and the liquid is pouring directly into that drain." You know that drainpipe leads directly to the city sewer system and eventually into the nearby body of water. It may even affect the local groundwater aquifer. You direct George to fix the problem immediately while you begin to prepare a report of the spill to the local agency. At your first opportunity, you inform the plant manager, Charlie, of the situation. He calls a meeting of you and George where the three of you discuss the situation. Charlie asks you how to take care of the situation without defaulting on the rush contract.

5.16.2 Discussion

Do you recommend reporting the leak even though you are concerned that it would risk stopping the production line and possibly cause your plant to delay the delivery of the order? Or do you recommend doing nothing, as George suggests, since the solvent is going right down the drain and the leak can be repaired after the order is delivered?

Regardless of where the sewer drain routes the toxic material, it is ultimately a threat to public health. An environmental professional cannot allow that to happen. Below is the recommended action for a trustworthy and responsible environmental professional who respects "doing the right thing." In dealing with Charlie, George, and your other colleagues in the plant, you must use care as well as justice and fairness to help ensure that this will be an important lesson for them on the importance of knowing the applicable environmental regulations and adhering to them. It is not just a matter of stopping or containing the leaking fluid but making the necessary tests to determine that the machines in the plant are sound and ready to be turned on anytime a new job order is received. It is not unusual for companies like ACME Machine to have a program for its manufacturing plants to have its machines serviced regularly, especially those with the potential to release toxic material.

5.16.3 Recommended Action

With the backing of the plant manager, Charlie, you must direct George and the factory staff to immediately install a collection container that will allow no further intrusion of the hazardous material to the environment. The collected material will be placed in a suitable container and labeled "Hazardous Waste." Next you must complete your report and take it to Charlie for his signature. Explain the importance of self-reporting if he shows any reluctance.

Then you must submit the report to the local enforcement agency. The enforcement agency will likely require that AMI's staff or a hired consultant characterize the degree of intrusion of the toxic waste and determine if a remediation will be required. You will be the primary overseer for your company in selecting the consultant, approving the work plan, following the investigation, reviewing the reports, and coordinating with the enforcement agency. The corporate legal staff will no doubt handle the legal action brought by the agency.

Finally, you must determine how the unreported leak occurred and why you were not notified immediately. You should then bring your findings to Charlie, who will handle any disciplinary action. Watch out because you may be one of the guilty parties for having not taken a positive emission or leak prevention program.

5.17 Reporting Previously Unpermitted Equipment in a Permit Renewal

5.17.1 Dilemma

As the plant environmental manager of ACME Production, Inc. (API), you are preparing an application to renew API's Title V permit. You discover that a coating unit was installed on the production line just before you came to work for your company two years ago. Your predecessor had not added it to the previous Title V permit. You investigate with the plant staff, and no one can shed any light on why this oversight occurred. It was simply forgotten.

The coater has been well maintained and looks brand new. No one from the local air agency has been to the plant in the past two years, and therefore the agency does not know that the coater exists. In completing the permitting for Title V, it will be necessary to apply for a local permit as well as to include it on the Title V. What installation date do you report for the coater on the permit applications?

5.17.2 Discussion

This dilemma involves whether or not to declare an omission by the person that you succeeded in a job that did not comply with a federal regulation. It deals with a situation where just because you might be able to get away with a minor infraction of regulations, it may not be the ethical way to handle a situation.

This may seem like an issue that could be easily ignored. It is tempting to just apply for a permit/license for the new coating unit using the current date, pay the license fee, and add it to your Title V renewal application. However, this is not the honest thing to do. It is clearly and reasonably not your personal or professional failure that the coater was never permitted. But it will be your failure to misrepresent the coating unit acquisition on the local district and Title V renewal applications. It is a good idea for any new environmental supervisor to review all of the plant's permits upon entering the plant and to reflect them in an equipment inventory to ensure that his or her plant is in full permit compliance.

5.17.3 Recommended Action

First go to API's plant manager and explain the circumstances. Explain that this is an opportunity to correct a minor permitting oversight by your predecessor. Point out all of your observations regarding whether the agency would ever be able to detect the true date. Then tell him that it could serve the company's best interest if API tells the agency of the true application and explains it as an oversight. Explain that it will build a level of trust with the agency so that API will not be harshly penalized in the event of a serious emission control malfunction. Suggest that the fact that an agency inspector has not been coming to the plant could be an indication that the plant has been in good stead with the agency. It would be desirable to maintain the agency confidence.

In preparing the permit application and the Title V renewal application, you will need to specify the true acquisition date. There are emissions from that unit that were not reported previously. Take those applications to the agency personally and explain the circumstances. If API's plant manager wishes to accompany you, accept his support. Work with the agency to rectify the situation. In a self-reporting situation like this, the agency may not exact a fine or, at worst, issue a first-offense type of penalty. Since you are new to the plant, this is a good opportunity to become friendly with the agency staff and gain their respect.

5.18 Resampling Discharge to Override Upset Condition Sampling Result

5.18.1 Dilemma

You are the environmental manager of a wastewater treatment plant in Pacific Grove, Oregon. Your treatment plant effluent is discharged into the ocean a mile offshore. Your staff has just collected a sample of the effluent from your plant. Janet, your plant's office supervisor, calls you and tells you that she learned of your staff's sampling, and that it had been taken when the plant was experiencing an upset. She asks you to resample because the one you took is not "representative" and the plant is operating normally now. How do you respond to her?

5.18.2 Discussion

This dilemma involves the environmental manager of a plant being requested to disregard an earlier sample and retest the discharge from the plant now that an earlier upset of the process has been corrected. As environmental manager of that plant you have a responsibility to protect public health. The environmental regulations, under which you operate, are designed to ensure that you will comply and thereby protect public health. In this case an upset has occurred and under the law must be reported. In this case, the quality of the effluent is checked by periodic sampling and analysis. If the sample analysis indicated that the effluent is outside of acceptable limits, the agency must be notified immediately and the plant has a specified time to bring the discharge into compliance.

5.18.3 Recommended Action

Since the plant effluent is now in compliance, you should explain to Janet that it would be appropriate to retest now, but that the results of the previous test must also be reported along with the cause, start time, extent of the upset, how often it has occurred in the past, and other details, along with the time of returning into compliance. Assuming that your test results confirm the out-of-compliance situation and the present return to compliance, you prepare a report and take it to Paul Smith, the plant manager.

Paul and you must decide on both the regulatory and disciplinary actions. If both wastewater samples are within acceptable limits, you may report only the one with the higher concentration of toxic waste. That would be the fortunate result. If the toxic concentration of the first sample is beyond allowable limits, you must submit a report to the water pollution agency explaining the cause and corrective action taken, as well as the amount of toxic waste released that was beyond the allowable limit. You should meet with the process department supervisor and determine her solution for ensuring that there is no repeat of this incident. Suggest a stream monitoring system that will immediately alert personnel of a potential violation and, if necessary, shut the process down until the cause can be detected and rectified.

Next, you must submit your findings to Paul. Help him to determine the corrective action to be taken and any disciplinary action necessary, if appropriate. In no case should you accept a decision by the department supervisor or plant manager to suppress the first-sample information. Finally with this documented information, submit a violation report to the local agency detailing the cause and the corrective action. This violation (assuming there was one) is a symptom of a potential weakness in the environmental control system. You need to take action and take the necessary steps to avert any further incidents like this one or any air, water, or hazardous waste discharge issues.

5.19 Boss's Subterfuge Puts You on the Spot

5.19.1 Dilemma

As an environmental professional you have been selected to be the environmental spokesperson for Evergreen Lumber Company (ELC), a forest products company. The company owns a huge tract of forestland that, according to ELC's long-term plan, will be harvested on a rotational basis over the next century. You are scheduled to meet with the local county environmental commission (CEC) to discuss ELC's long-term plans for the land. You are, in part, trying to dispel the rumor that ELC plans to open a landfill once the trees are cut down. The CEC approval is needed quickly because arrangements for harvesting the trees have been made and contracts have been let for its execution. ELC's owner, Ross Cloose, has charged you with the criticality of receiving CEC's approval because of his commitments involving the start of work.

As you sit back in your seat on the airplane, you open the folder on the tract of land. You realize that Ross's secretary mistakenly gave you Ross's folder on the land in place of yours. A detailed plan for the development of a landfill is in the boss's folder. How do you proceed? Consider your responsibility to your industry, yourself, the environmental commission, and general public health.

5.19.2 Discussion

Here you are all alone on the airplane, cut off from any source of communication with the office. Wild thoughts are going through your mind. Have I been set up by Mr. Cloose? Is the information in the file current? Since you have no one with which to interact, consider the pillars of character. Your boss has entrusted you as the company's representative. You must tell the truth and the whole truth. Do you trust the boss enough to believe that he would not set you up as appears in this situation? You should respect that delegated responsibility and believe that the folder information is out of date. It is your responsibility to determine the answer to these questions before meeting with the county commission? And what if one of the commissioners asks you about the rumored landfill?

5.19.3 Recommended Action

When you land, you must call Ross and explain the situation. He must explain the discrepancy or give you more information to reveal to the commissioners to ensure a just and fair finding by the commission. If you are uncomfortable with the information you receive from your boss, you have the responsibility to cancel your meeting and return to the office. In situations like this, we would all hope that the information in the folder is outdated. While this may have been considered as an option in the past, it has now been cancelled. You cannot go to the meeting until this is resolved.

And what if Ross tells you that the landfill idea has not been discarded? He says that it is not a matter that should be considered now and that a separate permit application would be submitted with full support of an environmental impact report if the concept was fully workable. You and he must discuss the details of the future of ELC's business plan. If you had known about this long-range plan, you would have cleared up ELC's intent before leaving the plant. But now that the issues are made clear by your call to Ross, you are adequately prepared to face the commission.

5.20 Big Boss's Friend Expected to Get the New Consulting Contract

5.20.1 Dilemma

You have been transferred to a new manufacturing division of Rincon, a Fortune 500 company, as its environmental, health, and safety (EH&S) officer. The division uses the same environmental consulting firm, Bright & Shiny Consultants (BSC), for most of its routine work. The managing partner of BSC, Bill Barton, and Rincon's division president, Charlie Jones, both belong to the same country club and are on the alumni board of a major university. You hear that BSC tends to do sloppy work, does not meet deadlines, and is expensive. While BSC has made no grievous errors yet, you get the feeling that it is only a matter of time until it does. A major plant renovation project is in the works, and Bill assumes that you will use his friend's consulting company to carry out the environmental permitting. How do you proceed?

5.20.2 Discussion

This is a difficult dilemma because you have no real evidence of a threat to public health, which should be your primary concern. If this division of Rincon is making a good profit and is in good standing with the community, it may well be the case that as division president Bill uses these kinds of political connections to enhance the plant's importance within the community. The corporate procurement policy may leave the issue of competitive bidding to the discretion of division presidents. However, this is not an EH&S responsibility. You may be convinced of BSC's poor performance. But as the new guy in town, you will not be able to convince Bill of this without some serious examples. Besides this issue, since you have just moved your family into the area and have your children in school here, any thoughts of getting on Bill's wrong side could have serious effects on your family. How should you proceed in this situation?

5.20.3 Recommended Action

What you should do is to engage BSC to do the permitting and hold its feet to the fire. It must be made to realize that you are the client and that your firm demands its respect. Meet BSC's project manager and attempt to build a friendly relationship with that person. Arrange for a meeting in which both of you meet the agency staff who will process your permit application. Later establish a personal relation with the lead person at the agency that would allow you to inquire about its confidence in the BSC staff. If the environmental agency has problems with its work, request a letter specifying those deficiencies.

In the meantime, you should observe BSC's work and ensure that it is complying with regulations and test procedures. You should be doing your best to protect public health by reeducating the consulting firm. Just be careful not to overlook the shortcomings in order to keep on Bill's good side. You may even enlist Bill's help in communications with BSC's managing partner. Often, top management is unaware of its employees' shortcomings until a serious issue arises. If you can solve the competency problem and maintain top management's comfort level, then you will have the best of both worlds. The more open and honest you are in this endeavor, the better your chances of achieving a good result for both Rincon and BSC. Just know that you are the responsible manager in this situation. You can be friends all around but you cannot forget your responsibilities. As the new guy in town, you must be keenly aware of that permitting effort, and when there are activities in which you are not expert, e.g., modeling or testing, do what is necessary to validate the results. You may learn that your predecessor did not do this type of supervision and that is why BSC did not do its best work.

5.21 Contract Term Prevents Consultant from Revealing Problem

5.21.1 Dilemma

The environmental consulting firm KGB that you manage has been dealing with a client, Advance Coatings, Inc. (ACI), that refuses to reveal a serious contamination condition that you are certain endangers public health to the local authorities. This client has had a seven-year working agreement with your firm and its fees have averaged $500,000 per year. The agreement with ACI has a confidentiality clause that prohibits your firm from revealing any information regarding your work for it. Because of the possible consequences to your firm, your project manager, Dean, has been secretly recording the phone conversations he has had with ACI's plant manager, Dan Routin. During the phone calls Dan has made veiled threats regarding any future business with his firm "if any word of this contamination issue gets out."

Your firm's attorney, Michelle, indicates that your contract with the client prevents you from taking any action to reveal the test data and indicates that Dean's recording of the conversations without notifying the client was against the law. How should you proceed?

5.21.2 Discussion

Clearly the objective here is to persuade the client to report the contamination action and to begin the remediation necessary to protect public health.

Contractually, you cannot reveal this incident to the authorities. The seven-year history that you have had with ACI suggests that this is not a usual circumstance. Based on the magnitude of your activity with this firm, it is apparent that it is a large firm in a substantial business. ACI's top management would not want a serious incident such as this to escalate into a high-profile incident. Dan evidently does not want to have the expense of this potential violation affect his plant's performance and profitability, which relates to his performance bonus.

Many large manufacturing firms do not charge the plant management with an "expense" to maintain compliance with environmental regulations. For example, if a scrubber fails, it is replaced but not charged against the plant manager's performance, profit-wise. Whether or not this is the case with ACI, it is a concept worth considering when communicating with Dan. Some ethical consulting firms refuse to agree to such a confidentiality arrangement when it comes to discharges that could seriously affect public health. Unfortunately in a competitive world, it often takes some ethical training of management personnel to respect that responsibility to protect public health. This applies not only to environmental professionals but to industry and corporate officials as well.

5.21.3 Recommended Action

You should meet with Dan in a friendly atmosphere, maybe over lunch. Discuss the issue off the record and explain the consequences that you both face if this condition is not reported. Explain that some of his employees are aware of the condition and that all it will take is a disgruntled employee to report the condition. Then you will both be in trouble. Suggest that there is a benefit to self-reporting. The positive benefit is to show concern to the agency to gain its confidence that you have the good intentions of correcting this condition. Finally, inquire whether ACI's corporate environmental director is aware of the incident. Explain how many companies charge compliance with environmental regulation to a corporate account. This meeting should result in an agreement for a plan of action. If it does not, KGB's attorney should contact ACI's corporate attorney. The bottom line is that the discharge must be reported and remediation action taken.

5.22 Client Believes You Overreacted

5.22.1 Dilemma

You are the leader of a consulting hazard operability (HAZOP) team assigned to evaluate the refrigeration system of a CO_2 (dry ice) plant owned by CO_2-R-Us, a national firm with plants in major cities around the country. The plants are located adjacent to refineries that provide the CO_2, which is liquefied and solidified by compression and refrigeration. Your consulting firm, EERI, has been contracted to perform a HAZOP analysis on several of the CO_2-R-Us plants. This is the first plant to which your team has been assigned. In the initial plant inspection, your team's structural specialist, Ken, finds a serious design flaw in the system that would cause a major release of anhydrous ammonia in the event of an earthquake. This plant is located in a maximum earthquake potential zone, and it experienced minor leaks in a previous earthquake. Ken is a registered structural engineer with extensive seismic analysis experience. He runs a routine calculation that convinces him that the design is definitely marginal, but there is nothing in the building code that covers the situation. You inform CO_2-R-Us' plant manager, Fred, of the situation and suggest structural reinforcements that Ken, you, and Fred acknowledge might cost in the range of \$300,000 to \$500,000. Fred is cold and somewhat hostile. He tells you and your team to leave while he considers the problem.

A week later, your supervisor at EERI, Leroy, who has been working with CO_2-R-Us for over 12 years, informs you that this contract has been terminated because CO_2-R-Us thinks that your team is reactionary. CO_2-R-Us's structural engineer, who worked on the design of that plant, says that there is no problem. You are stunned. You have Ken confirm the analysis. As a result, not only are you sure of the analysis, but you now have several case studies in which failures have occurred for similar reasons. You present this information to Leroy, who refuses to reopen the issue with the client. You are told that this client is not angry with EERI and will continue to use its services, but that, in this instance, there was bad chemistry between you and Fred and it was a better option to just end the situation before it got ugly. Your presentation of the latest findings does not change Leroy's decision. What should you do?

5.22.2 Discussion

This is clearly a sensitive situation. Certainly analysts can have differing opinions as to the integrity of a structure. In this case you have gone to the effort to validate Ken's initial analysis and have anecdotal evidence that this structure has a potentially serious problem. You have no information on the CO_2-R-Us structural engineer's analysis, but you do have faith in Ken's work. Fortunately, no serious earthquake has occurred in the plant area, so there are no actual data to draw upon. However, if Ken's analysis is reasonable, serious damage and possible injury and death could occur. In the interest of protecting public health, it cannot be ignored.

5.22.3 Recommended Action

As a trustworthy and responsible engineer, you should convince Leroy to allow you to submit a certified letter to the CO_2-R-Us corporate headquarters including information about how to strengthen its marginal components, and enclose a copy of your structural analysis. Also enclose any drawings you have made and send the letter to the client with "return receipt requested." You should also speak to EERI's corporate attorney and suggest that a letter be sent to CO_2-R-Us's legal department with the full disclosure of Ken's findings.

Since the earthquake failure mode may never happen, you may never know if you were correct. It is also wise to contact the city engineer's office and suggest that the code be looked into based on your findings. But sending the information to the client is the right thing to do. If you pose this dilemma to a group of structural engineers, you are certain to stimulate a lively discussion.

5.23 Your Client's Adversary Modeler Makes Error in Your Favor

5.23.1 Dilemma

The plaintiff claims that your client's plant, Standard Petroleum, Inc. (SPI), exposed the neighbors of the plant to a cancer risk that requires notification of the neighbors. The SPI plant is immediately upstream of a densely populated residential area. SPI has failed to make that notification and is being sued under a state law requiring notification. In your evaluation of the modeling you find that the building height and downdraft assumptions used by the other modeler were less conservative than you would have used. In fact, according to the "correct" assumptions, your client should have not only warned the neighbors, but also been required to install major emissions controls. When you reveal your findings to SPI's attorney, you receive a reply that the plaintiff has offered a financial settlement and will not pursue further remedies. The attorney decides to negotiate the monetary award and ignore the results of your finding. You inquire about the need to install controls. The attorney says that all of the modeling assumptions are so conservative that there is no need to go any further. You are told that your findings are work product and cannot be discovered. You are to erase the computer file and just submit an invoice without a written report. How do you deal with this situation?

5.23.2 Discussion

In this dilemma you are to evaluate a health risk computer modeling effort by your client's adversary in conjunction with a lawsuit. It is not unusual for attorneys and environmental professionals to have differing points of view on issues such as this one. The environmental professional is committed to protect public health, while the attorney is committed to resolve the case with the least impact on his client. Now that you have this finding, it would be ethically incorrect to ignore the analysis. Eventually the regulators will require an air toxic analysis of the plant's emissions and will require controls be installed to bring the emissions to acceptable levels. Even if the model that your staff has used is conservative, the effect on the population has a wide level of sensitivity. The acceptability levels have been established by the U.S. Environmental Protection Agency using comprehensive health effects studies. Conservative or not, if your modeling shows the level of exposure to carcinogenic emissions to be above the acceptable standards, then some action must be taken to protect public health.

5.23.3 Recommended Action

After the case is settled, you should go to your client and suggest that it would be in the company's best interest to determine whether or not the neighbors are being exposed to an excessive dose of carcinogens. If the neighbors begin to report cancer cases, they will open another case against the plant, and the lawsuits will begin again. You should suggest that the client retain you or another consultant to conduct an air sampling test to measure the actual carcinogenic exposure now. If indeed the cancer risk is high, then the plant can consider installing emissions controls to eliminate, or at least reduce, the exposure. If the tests show that the exposure is not excessive, then the client will have evidence in its files to counter any future attempts to sue it. If the concentration results are very low, the client might even want to share them with the surrounding residents and homeowners associations, etc., to convince them that the plant is really a good neighbor.

5.24 Industry Survey Data Point Eliminated to Fit Desired Profile

5.24.1 Dilemma

You are the elected chair of an industry committee that is negotiating a maximum achievable control technology (MACT) rule with the Environmental Protection Agency's Office of Air Quality Planning and Standards (EPA OAQPS). Your committee has gathered data to establish a MACT floor. A consultant's report demonstrates that "no control" is a MACT floor for a particular process. Approximately 500 sources have been surveyed and only

27 plants had controls. EPA OAQPS has accepted the report and is about to promulgate the rule.

You happen to attend an industry association meeting where you meet a colleague from another company that congratulates you for your accomplishment. "I was worried because I knew that ACME Corporation has over 15 processes on the West Coast and they're all under controls." That triggers a concern in your mind because you did not recall seeing ACME data in the consultant's report. You call the consultant, who tells you that several of your committee members told him not to bother contacting the West Coast firms. Their explanation to the consultant was that "the West Coast companies used a different process." You realize that a correct use of the data would require a control of all the processes. You call a committee meeting and tell them about the situation. You tell them that the agency should be advised immediately. The majority of them are opposed because of the huge costs of the control involved. They say that the industry could not stand the expense, etc. How do you proceed?

5.24.2 Discussion

The EPA regulation defines the MACT floor as a point between the current lowest level of control in the United States and the highest level anywhere in the country. Specifically, it is 82 percent of the range between the lowest and the highest emitters. All sources that are emitting more than the 82 percent point, the MACT floor, are required to bring their processes to at least the 82 percent level. Each industry—energy, petroleum, agriculture, etc.—is required to develop a MACT floor for each process in that industry. Over several years, industries formed committees and hired consultants who gathered data from the many sources in that industry and statistically processed those data to establish the MACT floor.

Because of the early discovery of smog-type air pollution on the West Coast, it was generally felt that the West Coast controls would represent the higher levels of the MACT distribution. Therefore, in this scenario, if the industry committee tried to rationalize that the West Coast process was different, and you know that this is not true, you need to at least determine the rationale that was used to separate the processes on the West Coast. You have confirmed the fact that the ACME Corporation data were not used. When you meet with the association committee, you move that the committee inform the EPA OAQPS of this oversight, and they refuse to support you. The issue in this dilemma is: Do you go along with the committee vote, or do you take action because their argument of excessive costs is invalid and you believe it to be less than honest?

5.24.3 Recommended Action

First, you go to your company management and try to gain its support. Assuming that it backs you, you or an executive from your firm go to the highest levels of the industry association. The decision might also be made to take this issue back to your original committee and have your new team try to convince the committee members to change their minds. It is likely that the EPA OAQPS would pick up on this deceit and embarrass your industry association. If this does not work, your company executive committee should do what is necessary to avoid this deceit. You need to help them where possible.

5.25 Agency Contract Modeler Uses Excessive Conservatism

5.25.1 Dilemma

You are chief of an air pollution control agency that has adopted a standard model that was prepared by a contractor. Your staff has performed trial runs of the model and accepted it based on the procedures prepared by the contractor. The model is released and industry groups and local agencies subsequently find it to overpredict emissions by factors of 25 to 100 percent. You appoint a select committee to evaluate the situation. They concur with the overprediction findings, but find that the value is really in the range of 10 to 25 percent. You confront the contractor, who indicates that your staff strongly suggested that it build in a conservative factor in order to put more pressure on industry. The contractor refuses to redo the model without charge and the price to correct the issue is significant. The agency budget is expended and the legal deadline has passed. How do you proceed?

5.25.2 Discussion

In developing health risk models such as this, often the agency staff overseeing the development of the models tends to use a conservative approach to protect public health. However, being too extreme places undue expense on industry and other emitters. Often industry committees work with the agency to seek a favorable compromise between public health and control costs. The consultant who developed the model has completed his contract and cannot be expected to redo the model without additional funding.

5.25.3 Recommended Action

As chief of the agency, you have clear evidence that the model is faulty and there are not funds available immediately to correct the model. Therefore, you must withdraw the model. You should confront the agency staff who made the decision to build in a conservative factor. You need to know why this was not elevated to you. You should convene an inquest of all of the involved agency staff and seek resolution of the issues. Following this you may need to discipline the people who made the arbitrary decisions without your concurrence.

5.26 Consultant Debates Bait-and-Switch Option to Win Bid

5.26.1 Dilemma

You are an individual owner of Blue Tape, a company that bids on an air permitting job that involved a substantial amount of computer modeling. As an expert with years of experience, you have the capability to perform the necessary modeling. But since you had several ongoing projects, you did not have the time to perform the modeling yourself. On this project, you had proposed to subcontract the actual computer work of modeling, but indicated that you would monitor the inputs and results carefully. The client chose a competitor that had less experience than your firm, but the modeler proposed for doing the work was an employee. In a debriefing, the client explained that it felt that modeling was the key element of the job, and that it believed the company with the in-house modeler would be more likely to stand behind the results than a contractor relying on an outside modeler. In the next bid, you contemplate stating that you would do the modeling personally and present your personal qualifications, which are superior to those of the other competitor's. When the job is awarded, you would hire a subcontractor to do the computer work. You would closely check the modeling methodology to be used, the format of the data reporting, and personally take charge of the conclusions, recommendations, and any presentation of those findings. Is this an ethical solution for the next competitive bid?

5.26.2 Discussion

A competent consulting firm must invest time in acquainting potential clients with the capabilities of that consulting firm and in learning the potential needs of the clients. The situation that the consultant in this dilemma found himself in is that he had to win the bid solely on the basis of the proposal, without knowing the evaluation criteria of the client.

5.26.3 Recommended Action

Do not expect the client to change its decision to award the contract to your competitor. The issue is closed. Learn from your mistake. If you knew that the client was concerned about the quality of the modeling and would trust it with your firm if you did the modeling personally, then you must do it for the client personally, or pass on the bid. To commit your time to the project and then delegate it to a subcontractor is unethical because it is dishonest, i.e., not trustworthy. If you know in advance that the job really requires your personal expertise but you cannot be available, then the respectful action is to approach the client, honestly explain the situation, and withdraw from the competition. If your previous relationship with this client is such that your modeling capability is appreciated, the client may find a way to adjust its schedule to accommodate your participation. If not, your honest approach and caring about doing the right thing for the client should improve your chances of getting a contract awarded to you in the future.

On the other hand, if the modeling required the use of a new program with which you had no personal experience, then it would not be unethical to locate a modeler with the required capability and arrange to hire that person. In the proposal you could describe the company's experience, including the experience of the new modeler. You would explain your role in the modeling process to reassure the client that it will have your expert oversight, and that you will be responsible for the final result. Honesty and fairness would require you to present your modeler's role in this aspect of the project, but ensure that the client could properly perceive the work of your team.

Finally, the fact that you lost the contract due to a specific whim of the client does not give you permission to cheat on a subsequent bid. You can use this loss to approach the client and explain the situation, showing the client that you understand its choice and that you look forward to future opportunities to work for the client. This kind of effort is an ethical marketing strategy.

6

Environmental Dilemmas—Judgment/Decision Situations

6.1 Company's Environmental Release—Where Does the Buck Stop?

6.1.1 Dilemma

You are the environmental manager for ACME Steel, a manufacturing plant located on the Monongahela River. You receive a call from a local marina manager who suggests that you look into whether or not your plant is the source of a disturbing substance that is floating in the water and staining the boats moored at his docks. You investigate and conclude that the substance is probably from the water curtain that collects overspray from your painting area. You confront Charlie, the painting foreman, who confirms the discharge. Apparently the runoff from the spray booth water curtain is routed to a wastewater treatment plant.

The pump at the treatment plant failed two months ago and was not repaired because the maintenance budget was expended for that year. Untreated water has been being discharged since then. Charlie's boss, Susan, the manufacturing manager, had warned Charlie that he would be fired if he exceeded his maintenance budget again. You speak to Susan, who is outraged, denies everything, and says the pump will be fixed within 48 hours. You know that the cost of the repair would adversely affect Susan's bonus, and you suspect that was the motive behind her warning. She directs you not to report the incident because "no one could ever prove it was from our plant." You know that there is a stiff penalty involved if the company is caught. You report your findings to your boss, Hans, the plant's general manager, who asks you for a recommendation. How could this happen? Should Hans fire the painting foreman or just impose time off without pay? What about Susan? Do you and she share in the responsibility for the release? What is the responsibility of the general manager, Hans, and of the corporation?

6.1.2 Discussion

This is a case of "where does the buck stop?" Clearly, the incident must be reported as soon as the facts are assembled, and with appropriate management and legal oversight. By employing proper self-reporting, punitive measures by the enforcement agency could be mitigated. The values involved here are primarily trustworthiness, respect, and responsibility.

As to the question of how this could happen, it appears to be a failure of management practice. It is common practice to relate compensation to performance in corporate management. This practice, while an incentive for good performance, can lead to improper practice, lack of training, and cover-ups. Firms with good environmental management systems in place will often exempt pollution control and other regulatory compliance expenses from their bonus incentive program. If this firm had this policy, chances are that this situation would never have occurred.

6.1.3 Recommended Action

Whether Charlie should be given time off without pay or possibly fired really depends on how he was instructed in regard to environmental compliance and other factors. If the foreman has a long record of satisfactory performance, is generally considered to be a good person, and really thought his job was being threatened, then a brief period of time off without pay is justified. This would be a wake-up call to him and other plant employees.

The manufacturing manager should be subject to harsher punishment if she failed to adequately instruct the foreman and, presumably, the entire shop staff on the importance of environmental compliance. Worse still, Susan violated the trust placed in her by corporate management, not only by polluting, but also by trying to cover it up. Depending on her years of satisfactory service and other performance factors, Susan should be demoted to a position where she is no longer in charge. She may even be fired.

Hans, the general manager, and you, the environmental manager, are not without guilt. Hans is responsible in the eyes of the law and, depending on how egregious the violation, could find himself guilty of a felony. You failed to install a proper environmental management program, including formal training with ethics workshops to train employees to do the right thing environmentally. Also, ACME Steel was at fault for not having a formal corporate environmental management system backed up with a strong environmental policy. They could have and should have instituted a policy exempting the local management's performance for costs to maintain environmental controls. Everyone involved in this violation should be penalized in some manner appropriate to their level of responsibility. It is suggested that an impartial arbitrator be appointed to resolve the various punishments.

6.1.4 Feedback

This dilemma is offered to stimulate further discussion among the readers. In a circumstance like the one portrayed above, there are many issues that must be considered and many values that must be tested before a final decision is made. By and large, when this dilemma was published in *EM* magazine, the readers advocated "doing the right thing," but took a more legalistic approach to solving the dilemmas rather than an ethical one. For example, one reader charged:

> The dilemma posed in the article has nothing to do with ethics! The environmental manager's legal responsibility is to report it to the agency. His responsibility as an employee is to get the company lawyers involved immediately! By electing to intentionally discharge untreated wastewater the manufacturing manager appears to have intentionally violated the Clean Water Act and is likely personally subject to criminal sanctions. By doing so she has exposed the company, the environmental manager, the general manager, and the top management of ACME Steel to potential criminal liability. Firing may be the most favorable outcome she faces. If this ends well, the company pays a fine, has to take preventive action under a consent decree, and only the manufacturing manager is convicted.

This comment is true, but people often have personal and family obligations that need to be considered in achieving the right thing with the least damage to the stakeholders. What if the manufacturing manager is a single mother? What if she consulted an attorney and was told that the agency could never prove who released the discharge? It is sometimes easier to take a hard view when "real" issues are not on the line. Working out a solution that protects the public and results in the least damage with the most positive outcome is what the six pillars are all about.

6.2 Consulting Firm Client Fails to Report or Remediate Site Soil Contamination

6.2.1 Dilemma

You are the vice president and partner of ACE Environmental Consulting, Inc. One of your clients, Plating-Is-Us (PIU), receives a phase II environmental site assessment (ESA) from your company reporting the presence of soil contamination involving toxic metals in three places on its 10-acre property. The report indicates that groundwater is threatened and recommends that the condition be reported to the appropriate authorities, and that remediation be initiated to prevent the contamination from reaching the aquifer.

Also, to the best of ACE's knowledge, the reported contamination could pose a threat to the water in the aquifer. But, without further investigation, it is not certain that this would adversely affect public health.

The report also contains a proposal for implementing the remediation. PIU does not respond to your company's proposal, and in following up, you learn that PIU has neither reported the contamination nor begun remediation. When your project manager, Sarah, reminds her PIU contact of the company's obligation to report the problem, she is told not to worry, it is being handled, and that PIU will notify ACE when it decides to perform the remediation. Sarah is also reminded that the purchase order for the phase II assessment contains a clause restricting the disclosure of any findings.

The issue comes into focus six months later, when another client, Cookies-R-Us (CRU), solicits your firm to conduct a phase I environmental site assessment (ESA) on a 5-acre property that is adjacent to the PIU plant. CRU is interested in buying this property to build a new bakery. Do you submit a proposal or politely reject it due to a conflict of interest? What is the correct approach and how should it be handled?

The investigation to determine the potential effects of the contamination would be expensive and could result in public disclosure of the situation. Both clients have given ACE their business in the past and can be expected to extend their patronage if ACE continues to gain their confidence. As a senior partner, you naturally wish to handle the situation to maintain both client relationships, but within the boundaries of ethics. But can you? Better yet, should you?

6.2.2 Discussion

You believe that the right thing is to discourage CRU from using this site. Your findings must not have been reported by PIU or there would be evidence of soil removal or processing on the site. Leaving an anonymous tip would eventually be tracked down to ACE. But refusal to perform the phase I analysis would raise suspicion.

Apart from a possible legal requirement to report this contamination, ACE has a primary obligation to protect public health. To be conservative, it must be assumed that a contaminated aquifer would constitute a health hazard for drinking water. Furthermore, this condition could make the adjacent property unsuitable for the siting of a bakery. ACE has a responsibility to see that the appropriate agency is notified and that a remedial investigation is initiated as soon as possible. ACE may have been negligent in allowing the matter to go on for so long without appropriate action. However, this obligation must be weighed against the obligation to maintain client confidentiality. The objective is to find a way to resolve the situation with due concern for the six pillars of character: trustworthiness, responsibility, respect, caring, justice, and fairness.

Three items commonly present in professional ethics code apply in this situation: "avoid conflicts of interest and disclose those known that cannot be avoided"; "keep information on the business affairs or technical processes of an employer or client in confidence while employed and later, as required by contract or applicable laws, until such information is properly released, provided such confidentiality conforms with legal requirements and other parts of the code"; and "hold paramount the health, safety, and welfare of the public, speaking out against abuses of the public interest that I may encounter in my professional activities as deemed appropriate per professional standards and existing laws and regulations." The primary responsibility of environmental professionals is to protect public health. This third item overrides all other considerations.

The question is, do you:

- Decline the job with CRU?
- Accept the job with CRU, but conceal the problem at PIU's plant?
- Accept the job and tell PIU that you will have to disclose the problem to CRU?

6.2.3 Recommended Action

The first step is to convene a meeting of your firm's management. The phase II ESA data must be reviewed thoroughly to ensure that they are sound. Next, the staff member who has the best relationship with PIU should be designated to approach the company about the situation. While CRU's interest in the adjacent property is immaterial with regard to the necessity for environmental reporting and cleanup, it would not be inappropriate to mention this to PIU as a motivating factor. PIU's failure to report the condition could be due to many things, including fear of prosecution, lack of remediation funds, or even being in the middle of a settlement with the agency that is not open to public knowledge. PIU may argue that the contamination was not a deliberate attempt to avoid the expense of disposal, and that the problem was in fact discovered due to the company's due diligence in contracting for the phase II ESA. ACE may be able to help substantiate this argument. A great deal of care and respect must be exercised in broaching the issue with PIU. If PIU's managers are unaware of the benefits of self-reporting, advise them to consult an environmental attorney.

Hopefully, PIU will agree with your assessment and the matter will become public knowledge. However, if PIU does not agree, then, after giving the company due warning, you must use appropriate legal means to report the finding. Your attorney must submit a notice of intentions to PIU and work through the procedure. The issue of breach of contract regarding the purchase order also must be addressed. It is possible that PIU could become vindictive. In most cases, after due consideration, the client will respond appropriately. But if this does not happen, your company should realize that this is not a client with whom it wishes to associate and should be prepared to handle the notification. Addressing the initial question of whether to submit a proposal to CRU, ACE is in the position of having knowledge that competing firms will not have with regard to the desirability of the adjacent property. In "holding paramount the health, safety, and welfare of the public," ACE has a responsibility to ensure that CRU is made aware of the situation. Until ACE's negotiations with PIU are complete, however, it would be a conflict of interest to submit a proposal to CRU.

Declining to bid at this time without revealing the reason would be an appropriate response. Once the situation is resolved (i.e., once the contamination has been reported), your company has an obligation to make CRU aware of the situation. Of course, CRU will be curious about your declining to bid. Suggest to the client that there may be some factors about the neighborhood that would not be favorable to its proposed operation, but that you are not at liberty to reveal them. This should cover your ethical responsibility until the other matter is settled. Tread lightly and use legal advice to mitigate the ramifications with PIU. It would be unethical for ACE to allow CRU to purchase the land without knowledge of the contamination and pending remediation. If you have established a level of trust with CRU from past dealings, then your suggestion will serve a useful purpose.

This dilemma has many innuendos worthy of discussion. For example, was it ethical for ACE to accept a contract that included a confidentiality clause if it felt that a condition affecting public health must be handled appropriately to protect public health? If it is the policy for a consultant to ensure that any spills or contamination are reported, shouldn't that be stated up front? Would such a statement included in the consultant's standard contract have a negative effect on business? These questions make stimulating subjects for an ethics workshop. It is suggested that you use this dilemma as a platform for periodic workshops to discuss this and other problems that arise in the workplace.

6.3 Agency Staff Suppresses Data to Justify Unreasonable Strict Standard

6.3.1 Dilemma

You are the executive officer of State Air Pollution Agency (SAPA). SAPA's board has just passed a new discharge standard for a pollutant based on extensive research on exposure and epidemiological data developed by your chief scientist, Jonathan. Jonathan personally conducted this research project for the past two years, devoting many weekends and evenings to the effort. One month after the promulgation of the new standard, a disgruntled subordinate of Jonathan's comes to you and confides that the scientist had suppressed certain data that indicate that the standard is unnecessarily severe. He says that Jonathan maintained strict isolation during his research and analysis, rejecting any opportunity for professional criticism. Also, he treated his coworkers with disdain by closing them out of the research phase and by believing that only he knew what was good for public health.

After reviewing the extensive industry testimony during the rule development phase, you realize that the severity of the standard will cost industry heavily. When you confront Jonathan, he admits to the accusation, saying that some of the data were spurious and that he felt justified in ignoring them, since, if anything, the public would be better protected as a result of enforcement of the new rule. Jonathan asks, "Are those industry reps whining again? I get so sick of their complaints. All they care about is money. But I am not going to let them run me." How do you proceed?

6.3.2 Discussion

In rule making it is sometimes difficult to recognize when a question of ethics is involved. Frequently, in the area of environmental management, a breach of ethics involves a practice that endangers public health and safety or tries to cover up a violation of a rule or regulation. Occasionally, however, a breach may involve a case of deviating on the opposite side. This might seem an unlikely scenario. How can someone be too honest or too caring or too professional? When this happened in an actual situation at the federal level, the national news media eventually learned of the incident and, after extensive investigation, reported the situation in detail.

One example is lying to save a life. Suppose you are standing on a street and a woman runs past you chased by two men. She screams, "They are trying to attack me!" as she dashes into the entry of a building around a corner. The men ask you, "Which way did she go?" What do you tell them? Clearly, the right thing is to lie. In this case the value of caring overrides the value of truthfulness. This exaggerated situation is used to illustrate that sometimes it is okay to violate certain values to protect public health and safety. In doing the right thing, ideally one should not have to make snap decisions and should take the time to investigate all of the facts (e.g., whether or not the woman was a thief and the men were police). Sometimes one must decide how much to sacrifice to ensure public health and safety. In establishing environmental regulations, the regulating agency must decide how safe and how stringent to make the regulations. For example, in the case of air toxic regulations, one standard may result in ten cancer cases per one million people. But why not set the standard at five or one or none? Who should decide?

When a proposed regulation will likely cause industry to close several plants in towns where they are the major employers, should the potential unemployment be taken into account? Can regulations go too far? Is it best for humanity to ensure that the environment be controlled to protect the most fragile individual? These issues will not be resolved in this book. But they are interesting, and the reader will need to inject this thinking into the evaluation of this dilemma.

First think of what is at stake here. The agency's credibility could be immeasurably challenged if this were to become public knowledge. Jonathan is a competent person, but in this case, his ego seems to have taken over his sense of reasoning and responsibility. But he was also so strongly committed to his task that he spent many hours of his own time to achieve what he thought was necessary to protect the public.

What values did Jonathan violate in this action? Three items commonly present in professional ethics code apply in this situation: First, he failed "Exercise honesty, objectivity, and diligence in the performance of all my professional duties and responsibilities." While he might have been diligent, he was neither objective nor honest. Perhaps the data he used were true, and therefore he felt that he had been honest. But honesty is more than just being truthful. It is telling the whole truth. For example, stating that a drug will cure a disease, but failing to reveal potential serious side effects, is truthful but dishonest. He also violated "Seek, accept, and offer honest professional criticism, properly credit others for their contributions, and never claim credit for work I have not done." He decided the result he wanted, selected the data he needed to prove his assumption, and rejected the contrary data. This is a severe breach of ethics. Finally, he violated "Treat coworkers, colleagues, and associates with respect." This was an egregious action that should not have been tolerated. The disciplinary action warranted by this action should be severe. Even if Jonathan's intentions were to protect the public, his actions were unacceptable and punishable. Because there are some complications in firing civil servants, disciplinary options may be limited. In this instance, he should either be strongly encouraged to resign or be demoted and a letter of censure placed in his personnel file. Serious effort should be taken in determining his punishment. As his supervisor, how would you handle the situation?

6.3.3 Recommended Action

First, conduct an evaluation of all of the data. To do so, assemble all of the data provided to the chief scientist and delegate a person or group to perform an expedited independent analysis of the information. The statistical validity of discarding certain data should be tested, and all accepted statistical protocols to validate Jonathan's conclusions must be examined. It is important that he not be condemned on the word of one disgruntled subordinate. Even if the scientist did admit to discarding data, he still maintains that it was the right thing to do. Show him due respect, at least initially. This should be due process. If it is found that the chief scientist's determination can be justified, the agency has the basis to defend its current position.

However, if the investigation proves that there is insufficient justification, corrective steps must be taken. It is important to make this determination as soon as possible, but without a reasonable measure of doubt. If the investigation confirms that the emission limits are excessive and a less severe requirement is appropriate, immediate action must be taken. The rule must be changed to reflect less severe limits. Jonathan must be disciplined. To minimize damage to the agency's credibility, it is recommended that a revised rule be issued with appropriate limits. It is important not to place the blame formally on the chief scientist. In fact, it is the agency's failure for not having verified his original decision. A mistake was found and corrective action was taken. Out of caring for an individual, the right thing is accomplished without belittling the chief scientist. But this does not mean that Jonathan is not to be disciplined.

6.4 Firm Has to Surface Competitor Deficiency without Bad-Mouthing

6.4.1 Dilemma

You are an environmental consultant that has been engaged by ACME Fabricators, Inc., a major manufacturing firm to execute a program required by a consent decree administered by a state agency. (A consent decree is a document issued by a court as part of a judgment in a case against a company that has violated an environmental law.) The program you conduct will determine whether or not ACME will have to install emissions control equipment, best available control technology (BACT), costing about $1 million. In the course of reviewing available information to prepare your plan, you review the reports of AAA Environmental, the source testing company that ACME has been using. You find that the test procedures used by AAA are substantially out of compliance with the EPA protocol for that test. You have heard of AAA and believe it has a reputation of being a low-cost provider and not of the best quality. However, the head of AAA is a former agency source tester with many years of experience and is respected by the agency. You note that the agency has accepted AAA's quarterly test reports without comment for over four years in spite of the fact that the protocol that is routinely presented in its reports does not comply with the EPA's. The ACME environmental staff has no stack testing experience and is satisfied with AAA's performance and its price. For this reason, you have engaged Superior Testing Co. to conduct the source testing for the consent decree program and suspect that Superior's results may be substantially different from AAA's.

When you submit Superior's protocol to the agency, the agency's test engineer rejects it because it is different from AAA's. It is apparent that the agency engineer has never bothered to compare AAA's protocol with the EPA's. When ACME's staff learns about this rejection, they are concerned that you chose the wrong firm in Superior and naturally begin doubting your competence. After all, AAA's work has been accepted by the agency for some time. You are concerned about the ethics of speaking negatively about AAA. In order to defend your choice of Superior, it might be necessary to speak negatively about a competitor, AAA.

6.4.2 Discussion

The problem here is that you do not know that AAA's results will differ significantly from those of Superior. Assuming that you are correct in regard to the protocol issue (i.e., AAA's procedures are substantially out of compliance with the EPA protocol), then you have the problem of convincing both ACME and the agency that you are justified in selecting Superior. It is evident to you that AAA's marketing skills exceed its technical skills. It has established a long-term relationship with both parties, and to date the parties have no reason to be dissatisfied with AAA's performance. However, because of the potential cost of the control equipment involved in the consent decree, it is important that the stack testing be performed competently and according to strict EPA protocol. Besides, the EPA will very likely have oversight on the BACT issue. If you do not trust AAA to do the testing, you are ethically justified in selecting another firm on a competitive basis. In this case, assume that Superior was selected on that basis (or you can clearly justify its selection on a technical basis).

6.4.3 Recommended Action

The first thing to do is to prepare a direct comparison of the EPA's protocol requirements with AAA's test procedure. Make it overtly clear that AAA's procedures do not comply with the protocol. It would be advisable to have an EPA staff member review the list of protocol requirements to ensure that it is complete and to determine how essential each step is, especially those that AAA has omitted or those from which AAA has deviated. Be careful to not identify ACME or AAA in this effort.

Next make a presentation to ACME. Invite the highest level of its management to attend. Explain that this program will be given careful review by the state, with oversight by the EPA, because of the BACT aspect of the issue, which is a federal requirement. Mention that the state agency is not familiar with the EPA protocol and has not reviewed ACME's previous test reports against the EPA protocol. Explain that you selected Superior Testing on a competitive basis to obtain an objective and readily defensible data set with which to evaluate the BACT requirements. Admit that AAA has served them well in the past, and since it had the state agency's approval in the past, it had every reason to continue to use AAA. However, now ACME is in the spotlight as a result of the consent decree. From now on, everything must be according to EPA protocol. Then explain how you will take this comparative information to the state agency to gain its acceptance of your test procedure that is in accordance with EPA protocol.

Above all, avoid alluding to AAA negatively. Take the positive approach that you are just using the best resource available due to the sensitive nature of ACME's situation. Take the same approach at the state agency by also arranging for a formal presentation there. Instead of comparing the requirements to AAA's procedure, compare the Superior procedure to the EPA requirements. Do not bring up the agency's previous rejection of the protocol, as presumably it was done by a junior-level engineer who will not speak up if he or she is now convinced that you are right. You should make points without even referring to the previous rejection. If someone from the agency brings up the previous reports from AAA, saying that another protocol was used, simply respond that he or she will have to discuss that issue with AAA. Whether the AAA data are more or less favorable regarding the BACT decision is immaterial.

This approach would meet the test of trustworthiness, responsibility, respect, and caring. It also complies with the A&WMA's code of ethics, item 12: "Seek, accept and offer honest professional criticism, properly credit others for their contributions and never claim credit for work I have not done." Your honesty in dealing directly with the situation should gain you the respect of your client and hopefully of the agency. By not bad-mouthing AAA, you show both caring and respect. You really do not understand the basis for AAA's source testing performance. It apparently knew what the agency would accept and has served its client (ACME) in the most cost-effective manner. If, as the result of your actions, the agency takes a more careful look at AAA's work in the future, you would have indirectly benefited public health, which is the first responsibility of an environmental professional. It is likely that the agency will not challenge the former work after it has been accepted by it. By opening ACME's eyes you should gain its respect. Ultimately, it would bear the brunt of an eventual discovery of AAA procedural failure.

Finally, it is not unlikely that you will have to face AAA's manager. You should use diplomacy. Try to go over the list of noncompliance areas objectively. Say things such as you know how busy it is, and keeping up with changes in EPA protocols can sometimes be deferred. Hopefully AAA will use this as a learning experience and take it as an opportunity to upgrade its company policies.

6.5 Recruiter Wants You to Bring Your Customers and Key Staff

6.5.1 Dilemma

You are the regional manager for Reliable Environmental Solutions (RES), a large, well-known environmental consulting firm. You have been with RES for 11 years, and during that time you have developed the business in the greater Houston area. Under your direction, RES has been extremely successful and your superiors have recognized your value by increasing your pay, benefits, and other perks. Your lifestyle has steadily improved, and you enjoy the benefits of travel and expense account living.

One day, while attending a national trade show in Baltimore, where you are exhibiting your firm's capabilities, you are approached by Global Environmental Management (GEM), a smaller competitor, with an offer to become the company's vice president of operations at a guaranteed 50 percent increase in salary and the potential to double your income if the company's profit goals are achieved. GEM makes it clear that you are expected to bring with you the customer contacts, technical and regulatory training programs, and certain key personnel at RES that you have developed over the years. GEM wants you to start as soon as possible, offers you a two-year employment agreement, and even offers to take out a performance bond to guarantee that agreement. The company also offers to pay for all relocation expenses, including the sale of your home and any other incidentals.

The additional income would be very nice for your family, whose social status and financial needs are steadily increasing, especially with children approaching college age. How do you respond?

6.5.2 Discussion

This offer is clearly a boost for your ego and would be a significant advance in your career. Do you accept on the spot before GEM changes its mind? Or do you reject it out of hand, thanking the company for recognizing your value? Neither. Decisions that are made in haste reflect your inherent selfish instincts. You need time to think it over carefully. Clearly, GEM has done its due diligence in researching your qualifications. As a competitor in the business, GEM has determined your value in advancing its business objectives and will want to get your commitment as soon as possible, before other considerations have time to dampen your natural excitement. However, you need time to get your ego under control and make sure that you are doing the right thing. Thank GEM for its attractive offer, but ask for time to evaluate your decision. The company will understand. If not, then you should be suspicious of the offer.

6.5.3 Recommended Action

Consider the stakeholders involved: yourself (career growth and prestige), your family, your current employer (RES), and GEM. First, consult your family. The new position is in another city; assume it is a significant distance from where you currently live. What does this mean for your family? An important ethical value to be considered in making your decision is caring. Your spouse's work may be an issue. Your family connections in the current location may be strong and dutiful. For example, there may be elderly parents or close relatives to consider. Your children may be unwilling to leave their schools and friends. Approach the family in a tactful manner, beginning with your spouse. If that goes well and he or she is willing to move, then ask your children if they would like living in the new city. Make it more like a discussion of where to take the next family vacation, so that it does not appear threatening. You have many more factors to consider before you make a decision, and it would be less threatening if the early discussion is exploratory and whimsical. When your decision to move is more solid, then you can take a more definite approach.

Another issue is your time with the family. A senior executive's schedule is not nine to five. You may find that the new job means more traveling and spending more time away from home, as well as spending more time at the office on managing, planning, and marketing. If, for example, coaching your children's athletics teams is a priority in your life, make sure that you are comfortable with the additional burden of the GEM position. The job offer may be a great opportunity for you, but you have a responsibility to respect the feelings of your family and be fair in allowing them to express their opinions in making this decision.

Assuming you are comfortable with proceeding at this point, how do you approach your current employer and what are your ethical obligations to your company? GEM expects you to bring with you the customer contacts, technical and regulatory training programs, and certain key people from your present firm. While this is not an unusual request, it does relate to the value of trustworthiness. As a loyal employee of RES, how much of its intellectual property can you ethically take with you? Because this issue involves more than just ethics, it is suggested that you consult an attorney. It may be illegal for you to take with you proprietary designs, computer models, and other software. However, your know-how to establish regulatory and technical training programs that you developed at RES for in-house use is different. It is considered personal property that you are entitled to take with you and use. An attorney may disagree, however.

The National Society of Professional Engineers (NSPE) code of ethics states: "Engineers' designs, data, records, and notes referring exclusively to an employer's work are the employer's property. The employer should indemnify the engineer for use of the information for any purpose other than the original purpose." The environmental equivalent of designs, data, and records would be proprietary hardware or software.

Customer information (i.e., names and contact information of former clients and colleagues) is your property. You know your clients, and if they trust you personally, they are free to choose to work with you in the future. For example, when a salesman leaves one firm and joins another, his list of contacts goes too. But any contracts you hold at RES must stay at RES. It would be unethical to persuade a client to cancel an existing contract with RES and award it to GEM so that you can manage it. Persuading a client to drop RES would be like stealing an item of valuable property.

As far as taking key people with you, it would be inappropriate for you to actively recruit your subordinates or colleagues for GEM. However, if those people should approach you regarding employment at GEM after you have left RES, it would be entirely ethical for you to evaluate their capabilities and offer them a position at GEM, if you thought their skills matched GEM's needs. Until you have made a decision to leave, it would be irresponsible, unfair, disrespectful, and generally unwise to have any discussion with other RES employees regarding GEM. Again, consulting the NSPE code of ethics for reference, the most appropriate guidelines are: (1) "Engineers [i.e., an engineering company] shall not, without the consent of all interested parties, promote or arrange for new employment or practice in connection with a specific project for which the [recruited] engineer has gained particular and specialized knowledge [presumably from work with another firm]" and (2) "Engineers [i.e., an engineering company] shall not attempt to attract an engineer from another employer by false or misleading pretenses." The first statement implies that, according to the NSPE code of ethics, GEM was unethical in making you an offer without first clearing it with RES. Although this sort of practice is not uncommon in industry, most managers recognize that it is important not to stand in the way of employees' advancement, even if it means losing them to another company. Conversely, managers are encouraged to promote talented people, so that they will be content to remain with the company.

Another issue of concern is employment or "noncompete" agreements. Many consulting firms protect themselves by having employees sign an agreement preventing them from working for a competitor. This is often included in the general agreements that one signs on entering a new firm. Before accepting the offer from GEM, you should check with the human resources department at RES to determine what, if any, agreements you may have signed. Also, be wary of GEM's employee agreement and make sure that it is acceptable. Assuming that you are comfortable with GEM's offer, your family's acceptance of it, and you are prepared to leave RES, it is time to notify your employer of your intentions. Be sure you are ready to leave because you may be terminated precipitously. Assuming that your relationship with RES has been excellent, your notification is likely to create initial disappointment. It is important that you present the notion with respect, honesty, and care. Explain how difficult it has been to make the decision and how happy you have been at RES. Explain that when opportunity strikes, you must take advantage of it. Talk about how the two companies may work together on joint projects in the future. But above all, be prepared for an angry response. When anger arises, it is best to excuse yourself. Understand that you have bruised your employer's ego.

You may receive a counteroffer from RES. Give it the same consideration you have given the one from GEM. Keep in mind that if you decide to stay, you may have to live down the incident. Some upper-level managers may feel that you used the GEM offer to extort a promotion and a raise from them. It would be advisable, therefore, to confront the RES managers with your concerns.

In an ethically ideal world, GEM's president would have contacted RES's president to ask for permission to make you an offer. Given that we do not live in an ethically ideal world, whenever you are faced with a job change opportunity that has ethical implications, be certain to consider and follow the five pillars of ethical values: trustworthiness, respect, responsibility, justice and fairness, and caring.

6.5.4 Feedback

After the above discussion was published, the following commentary was received:

> This morning I read [the] article on executive recruiting. I enjoyed the material and found it very interesting. While I agree with you that taking clients and documents with you when you leave a firm is, in most cases, unethical, I was surprised to see you say that, in an ideal world, a company would ask permission before making an offer to another company's employee. Really; what company would grant permission, and what company would ask? Business seems so cutthroat that I really cannot imagine that actually happening in real life.
>
> In an age where companies lay people off on a whim, force them into early retirement, and take away benefits every year while only offering minimal raises, I have a hard time developing any loyalty. I am 36 and still somewhat young in my career, but I honestly believe that if I get five years at a time from a company, I am lucky. What with mergers, bankruptcy, buyouts, reorganizations, etc., the last thing on company executives' minds is my well-being. I am a firm believer that I have to look out for myself because no one else will. Folks around my age and career path typically view jumping to another company as the only real way to get ahead. I hate to be cynical, and I hope you can tell me how wrong I am, but if I were to get an offer to leave my current employment, and it was a good move for my family and career, I would go without an ounce of guilt. Granted, I would give proper notice and carefully decide what should go with me, but I would go nonetheless. While the management at my firm seems to appreciate my contributions, the bottom line is that I am still just a number in a huge corporation and one small business move tomorrow could put me out of work.

The ethics of job change is to make certain to consider all of the stakeholders. If you are married with children, it would be unkind to make a precipitous move that involves a major relocation without consulting with them first. It seems that today many companies treat employees as commodities. An ethical employer would never treat employees in this way. An outstanding employee that could not be promoted and who was offered a job with another firm would be released with encouragement and wishes of good luck.

Furthermore, a supervisor who does not recognize the need for good employees to grow within the company will quickly lose them to another firm.

Another reader, the CEO of a national consulting firm, sent the following response:

> We have always felt that the common-law principle of the "master–servant" relationship applied to our staff. The idea is that when an employee changes jobs, he or she is obliged to do nothing to intentionally harm the former employer. In the area of marketing, this generally means that one should not contact the clients of a former employer for at least 18 months. In addition, there is the issue of whether your new employer wants you for your book of business or your talent. Unfortunately, too many firms only see a book of business, not the inherent skill of an individual to develop it. Thus, when switching jobs, it is wise to understand what the new employer has that offers greater opportunity to achieve results.

The authors were unaware of the common-law principal regarding an 18-month hiatus in contacting the clients of a former employer. We accept it as a reasonable guideline for doing the right thing. We just wonder why none of the codes of ethics include it. But this CEO's advice of understanding "what the new employer has that offers greater opportunity" is sound.

Later that same CEO offered the following comments on executive recruiting in general:

> The environmental industry is currently experiencing a good deal of "predatory" staff recruiting, so this discussion of ethics of executive recruiting is particularly timely. My comments deal with the law, how to change jobs, and perhaps most important, defining career goals prior to considering a job offer.

The most important point to be made is that state law governs employer-employee relationships. Approximately 10 to 15 states have sought to write specific laws that address this relationship, but the majority of states rely on old common-law principles and the so-called master-servant relationship. Under this concept, a servant who goes to work for a new master can do nothing intentionally to harm the previous master. In other words, an employee is free to choose to go to work for a new employer, but he or she is not allowed to take anything that was developed while on the payroll of the former employer, including computer programs, client lists, employee lists, and trade secrets (trade secrets may have to do with how to bid for work or how to achieve engineering results more economically). This concept is the basis for the secrecy and noncompete agreements that are common today.

6.6 Governmental Official Offers Funding Help in Return for Start-Up Partnership

6.6.1 Dilemma

You are the owner of Ortho Research Company (ORC), a small research activity that has been awarded a federal government grant available to small businesses on a competitive basis. ORC is contracted to develop an instrument to measure low concentrations of a chemical known to be lethal in high concentrations and which may have potential health impacts even at concentrations below current detection levels. At the end of the initial phase of the contract, your firm has built a laboratory-scale instrument that exhibits potential for commercial development. You submit a follow-on proposal for developing a prototype instrument. You contact Raleigh, the civil service official serving as your contract monitor, for instructions on applying for a patent for this technology. Raleigh notifies you that he will visit your plant for a site inspection and to discuss the follow-on effort in person. During the lab tour and demonstration, Raleigh shows great interest in the instrument. After the demonstration, you adjourn to your private office for a discussion. Raleigh tells you that you are a great engineer, but obviously inexperienced in business. He suggests that you incorporate your firm and that he become a silent partner with one-third ownership. Raleigh says he will see that you get the funding needed to bring the instrument to market and minimize obligations to the government. You are shocked by the proposal and react nervously to the idea saying, "I need to give this some thought." Raleigh responds by saying, "OK, this conversation never happened. If it ever gets out, your program will be terminated. I warn you that you better decide before the follow-on proposal expires." How do you proceed?

6.6.2 Discussion

This dilemma emerged from a discussion about marketing environmental technology at a session held during an environmental conference. This problem is a tough one, and it is suggested that the reader give it serious consideration.

Obviously, this is a form of extortion. It's akin to a shopkeeper being offered "protection" by a mobster. In your state of shock, the proposal might appear to involve the making or breaking of your company. On one hand, if Raleigh gives a negative report and recommends against awarding your company a follow-on contract, then all of your work to date could be compromised. On the other hand, if he gets behind your company and gives your work high ratings, your firm could have guaranteed success. Clearly, the latter approach is wrong both morally and ethically. So, how does one do the right thing with the least amount of damage? Whenever your instincts tell you that there is something disconcerting in a situation like this, take time to think things through carefully and seek help. Discuss the situation with a colleague, trusted friend, or independent advisor. Considering all of the issues involved, as discussed below, may help you arrive at the right solution.

6.6.3 Recommended Action

Immediately after Raleigh departs, document the happenings in an internal log or memo. Record all times and places of the visit, and interview your staff to extract any additional details. If someone recalls Raleigh being in your office (i.e., where the conversation took place), record his or her recollection of events. Determine if any of the conversation might have been overheard. Be careful not to "suggest" any of the topics discussed. This document should provide as detailed a recording as possible of the verbal exchange. Maintain it in your company files. A few days later, approach Raleigh and reveal your concerns that his proposal is dishonest. Be candid. Say, "What would happen if it were discovered that you had part ownership in my company? You would be found to have a conflict of interest that would compromise both of us. We would be taking a great risk on a laboratory-based concept that has yet to be proved beyond that point." In this discussion, show the official some respect. Appeal to his sense of responsibility by saying, "In your position, I believe that you would want to avoid any appearance of impropriety. Please give this some more consideration. You have an important job in the government that would be compromised if this were discovered." Make this discussion about Raleigh, not about ORC.

Say, "I have lost sleep over this. My conscience will not allow me to go along with this. I know that if you give this whole idea more thought, you will reach the same conclusion." If your appeal to Raleigh's conscience does not bring an end to the dilemma, and he continues to pursue the notion of becoming a silent partner in your company, it is time to take further action. Start by reviewing the grant contract regarding patents. Contact the grant office to initiate a patent request. You may learn that in these early development programs it is advisable to patent the concept for an invention before applying for a grant. Most grant programs, such as federal Small Business Independent Research (SBIR) programs, encourage patents, and ORC has the right to use the patented technology in the commercial market. However, the government also retains a limited right to use the technology.

At this point, it would be prudent to consult a patent attorney. Inform the attorney of the situation and provide him with a copy of the documentation (including your record of Raleigh's visit). Then proceed according to the attorney's instructions. In the event that you do not contact an attorney, the following action is suggested: contact the patent section of the grant office and let its staff know that you intend to submit an application. Presumably, you have official reports that were submitted documenting the success of the laboratory-scale instrument. Ask for instructions and determine whether you can submit the application directly to the patent staff office. It would be advisable to determine the role of the contract monitor in the patent process. It is likely that in the SBIR program, the technical contract monitor has nothing to do with the patent process, which is handled out of the SBIR office. This may not be true for other grant programs, or other jurisdictions, but it will be important to bypass the contract monitor to fully understand your rights. In communicating with the patent staff, if possible, avoid mentioning the name of your monitor. With regard to the follow-on contract, if you need to pursue the status of the proposal, contact the contract office directly. Maintain communication with that office and do not include Raleigh on the distribution. If Raleigh's name comes up, do not suggest any impropriety. If an opportunity arises, ask whether the follow-on contract will have the same monitor. If questioned, suggest that there were some technical issues on which you differed and another monitor might be more compatible. There is no reason to escalate the issue if it disappears. If ORC's stand on the matter resulted in the official having second thoughts and backing off, then it should end there.

On the other hand, if Raleigh continues to pursue the issue, and you have taken the actions noted above, warn him that you are prepared to take the matter to a higher level. Indicate that you have documented the incident, consulted an attorney, and are willing take legal action if necessary. If that does not stop the intimidation, then proceed by taking legal action. It is your civic duty to see that the official does not do the same thing to other unsuspecting companies.

6.7 One Firm Wants to Report Another Firm's Exceedance to Protect Public Health

6.7.1 Dilemma

You are the environmental manager at A-Chem chemical plant, located in a mixed residential and industrial area. Adjacent to your chemical plant is B-Oil oil refinery. Adjacent to both industrial facilities is a residential neighborhood. Your knowledge of B-Oil is fairly extensive, since the two facilities purchase a number of goods and services from one another and have done so for many years. A significant percentage of the gas fuel used by A-Chem is supplied by B-Oil. B-Oil consumes approximately 80 percent of the gas fuel it produces and sells 20 percent to A-Chem.

At 5:00 a.m. one Sunday morning, a malfunction occurs at B-Oil that causes the fuel gas to go off-spec for ammonia. Significant odors are immediately detected coming from all operations at the refinery. Operators working at A-Chem also notice the odors. Despite repeated calls to B-Oil, the A-Chem operators are unable to get any information on the extent of problem. At 6:00 a.m., B-Oil management issues a "shelter in place" warning for your facility and the adjacent residential neighborhood. You are notified at home of the warning that has been issued. You call your operations manager and recommend that he switch from using the purchased B-Oil fuel gas to pipeline natural gas. In addition, based on your belief that excess ammonia levels may have been present in the fuel gas, you advise that a precautionary reportable release notification be sent to the National Response Center for ammonia and nitrogen oxides. You also request that a sample be taken of the B-Oil fuel gas and analyzed for ammonia.

Later that day, you are informed that the analytical results indicate that while the nitrogen compound level in the fuel gas was higher than usual, it did not exceed a reportable level of any nitrogen compounds. Based on your knowledge of B-Oil, your expert opinion is that the B-Oil facility certainly did exceed at least one reportable quantity and possibly several. In fact, you have a strong suspicion that short-term health exposure levels were exceeded in the nearby residential neighborhood. However, during the following day's news coverage of the incident, a B-Oil spokesperson states with certainty that the highest levels of exposure occurred inside B-Oil's fence line and that the reportable quantities were not exceeded in the nearby residential neighborhood. Your immediate response is to determine whether you have a legal obligation to correct the misinformation. Your attorneys assure you that you do not. Your ethical dilemma is whether or not to correct the misinformation given by B-Oil. As an environmental professional, a person who has an obligation to protect human health and the environment, what is your personal and professional responsibility in this case?

6.7.2 Discussion

Clearly, the key ethical value here is responsibility. If you believe that the public is at risk as a result of this incident, then you must notify the appropriate authorities without delay. However, it is important not to diminish B-Oil's reputation and possibly involve the firm in a felony case, unless you are certain that this egregious condition actually existed. You have extrapolated your test data to estimate the probable B-Oil ammonia release, and therefore you must be honest with all of the stakeholders in this incident.

First, you need to convince A-Chem's management that they may have an ethical responsibility to notify the public. Your company attorneys' response may be correct in that A-Chem would not be committing a felony by not publicly correcting the misinformation. However, if some residents were to become ill as a result of the refinery's malfunction, and it is eventually determined that your company had cause to warn the public, then A-Chem could be named along with B-Oil in a personal injury suit. Second, it is imperative that you verify whether or not B-Oil exceeded the reportable emissions. You must convince the A-Chem manager with the closest contact at B-Oil to hold a meeting during which you explain your hypothesis and reveal your data. You should also request that B-Oil reveal the basis for its public statement.

While your motivation as an environmental professional is to protect public health, management may or may not share that obligation. However, most managers want their companies to be good neighbors. Failing that, no manager wants to be involved in a lawsuit brought by the firm's neighbors.

6.7.3 Recommended Action

You must be prepared to present your strong feelings on this matter and suggest (not threaten) that if it gets out that a cover-up has occurred, there is likely to be bad press at a minimum, and possibly even major penalties. It may be that (and you should hope for all concerned that this is true) your hypothesis is wrong and B-Oil has credible evidence that the reportable levels were not exceeded during the malfunction. Assuming this happens, you must be prepared to apologize to everyone involved and explain that your personal code of ethics did not allow you to dismiss what you truly believed was a potential threat to the local community. Further, explain how relieved you are to know that you were wrong. If you have maintained the respect of your management to date, this treatment of a potential problem should not be a black mark against you. On the contrary, your respectability and trustworthiness should be enhanced in their mind, since you took some personal risk to do the right thing. If this perfect-world scenario does not work out, document the transaction in a memo and send it to your immediate supervisor. Then, start looking for another job. If there were no public health problems as a result of the incident, it would be imprudent for you to attempt to publicly correct B-Oil's misinformation. The members of B-Oil's management will have to live with their consciences. It would certainly be a different matter if serious illness occurred as a result of the incident.

As an environmental professional, the responsibility to protect public health is your number one priority, but it has to be somewhat tempered by the circumstances of a given situation. To have simply dismissed your concern when you were told that you had no legal responsibility would have been a serious ethics violation.

6.8 Client Attorney Directs You to Disregard Your Additional Contrary Findings

6.8.1 Dilemma

You are a principal engineer specializing in risk assessment working for All-Purpose Environmental (APE), a national consulting firm with 27 offices located throughout the United States. APE has been retained by a prominent law firm, Win-or-Else (WOE), to support a case in which a client of WOE is suing an oil refinery for causing the client's cancer. You have been assigned by WOE as project manager because of your experience and knowledge of the petrochemical industry, especially toxic pollutant emissions. Your health risk investigation involves reviewing information on the refinery's processes published in a peer-review journal and analyzing emission factor data from the U.S. Environmental Protection Agency and other sources, on which you base an emission estimate to use with the health risk assessment model. To provide further validation of the exposure assessment, you monitor the ambient air at the refinery fence line and the client's residence for five days.

Based on the estimated emissions and the meteorological data provided by the local air district that has a monitoring station located near the refinery, the cancer risk you compute is 26 in 1 million, well above the allowable risk for new sources. The ambient test data, on the other hand, show a level of toxic pollutants at or below the detectable limits that is two orders of magnitude lower than the levels predicted by your risk model at the corresponding locations. You repeat the ambient measurements at times when the wind direction is moving directly from the refinery toward the residence's home and the refinery fence line and verify that there have been no recent outages at the refinery. You are satisfied that the air measurements are reliable, and so you present the data to the client's attorney, Clint at WOE.

After a period of extensive grilling and rationalizing, Clint directs you to prepare a comprehensive report of your research findings using emission factors and modeling. In addition, however, Clint requests that you disregard the ambient air monitoring findings and not divulge them to anyone. His arguments are (1) emission factor data are often used as conservative estimates of emissions to compute health risk for regulatory purposes in determining whether or not a new source can be permitted, and therefore sufficient evidence to present in this plaintiff's case, and (2) the health risk assessment model is based on long-term exposure to the pollutants, and WOE's client has lived in this residence for 30-some years, while the ambient air monitoring data are just a snapshot in time (measured for only five days) and will therefore only confuse the issue. Had the monitoring data showed greater exposure to the toxic pollutants, Clint might have considered introducing them.

You ask Clint if you will be subpoenaed to give evidence in this case and what you will say if asked whether you did any additional work to determine emissions. You are told not to worry, that WOE will prevent you from being subpoenaed. How do you deal with this situation?

6.8.2 Discussion

When doing research and performing data analyses in support of litigation action, an environmental professional may be forced to work with a less than objective client. Litigation is an all-win venue. Indeed, the notion of objective research might be compromised when the research findings do not support the adversarial objectives of the client and attorney. This dilemma explores the issues when an environmental professional's research findings disappoint the client's attorney.

Although your conscience may bother you (and indeed it should), you need to give careful consideration to this issue before deciding how to deal with it. WOE hired APE to do a specific job. Therefore, WOE owns the results of your work and can suppress it as long as they respect the contract and pay you for the work. By completing the health risk assessment report, you will have fulfilled your responsibility to your client. The question is whether the ethics of justice and fairness dictate that you have a responsibility to the defense in this case, and whether that trumps the need for loyalty to your client?

As a responsible person and an environmental professional, there is only one thing that comes before your family and career, and that is a threat to public health. But there is no public health issue involved here. If anything, your data seem to indicate that the health risk is below a level of concern. There are no issues of family or career involved either, assuming you continue to do a thorough job of documenting your findings and you do not take any precipitous actions that might anger WOE or APE. Thus, the matter becomes one of contractual commitment. When supporting litigation, it must be understood that each side is in a serious adversarial engagement. Each side is obligated to do whatever is ethically acceptable to win the case, and there is no obligation to help the other side.

When an ethical concern arises in your practice, do not react without first giving it serious thought. In a normal situation you might seek consultation with a supervisor, colleague, or friend on the matter. But in this situation you are committed to maintaining the confidentiality of your findings, revealing them only to your client, WOE. In this case, you cannot discuss your findings with APE management before presenting them to WOE. You should not have been surprised when WOE directed you to suppress the ambient data. When archiving this kind of sensitive data, be sure to place a clear warning that the file is confidential, that it is the property of WOE and can only be released to that law firm.

6.8.3 Recommended Action

Proceed as directed by Clint. If you are subpoenaed to give evidence in this case, you cannot lie. But you must limit your testimony to the modeling results or to whatever areas Clint instructs you to cover. In legal testimony, a witness must testify only on what he or she is asked. If the examining attorney asks if you computed the risk, the answer is yes. If asked how it was computed, you explain how. If asked if you used any other means to determine risk, Clint may instruct you not to answer, and you may say so to the court. If the court orders you to answer, over the objection of your counsel, then you must tell the truth. But if the examining attorney is unaware of the test results and fails to ask about them, then you have no ethical responsibility to disclose this information to the court in this case. Because of the likelihood of expert witnesses being tripped up by cross-examination, attorneys often rehearse the deposition with the expert witness beforehand. This is a wise procedure. Furthermore, if the opposing attorney poses a question that makes you uncomfortable, your attorney should be alert enough to intervene and object.

It may appear that you are being involved in some type of deceit by suppressing the ambient data. This is not an issue of an undisclosed threat to public health. The modeling results are the more threatening indication of a health threat.

It is important to understand the client's objective and to work together to meet that objective. Often an attorney's priorities will differ from those of an environmental professional. This case is one in which no ethical problem is involved in complying with the data suppression directive given by the client's attorney.

6.9 Client–Attorney Privilege—Exempts Only Imminent Threat of Harm to Persons

6.9.1 Dilemma

You are an environmental professional and you conduct a phase I environmental site assessment (ESA) in conjunction with the transfer of a property that is the site of a former steel manufacturing plant. The plant's operation closed down 40 years ago. You are engaged through escrow to conduct an impartial ESA for both parties. So far you have been unable to find any documentation of environmental impacts from the former steel plant, other than it occupied the location. At the time of the plant's closure, there were no regulations requiring environmental controls or the preservation of any relevant records. The phase I ESA doses not involve any testing (i.e., soil sampling, etc.). You have experience with steel manufacturing plants with regard to issues related to the Comprehensive Environmental Response, Compensation, and Liability Act (CERCLA) and the Resource Conservation and Recovery Act (RCRA), and you firmly believe that the property is contaminated. You state your beliefs in your ESA report. Escrow is closing tomorrow, and the seller's attorney, a regular client of your firm, calls to ask you to "lighten up" the language in your report and take out any references to historical problems associated with former steel plants. How do you proceed?

6.9.2 Discussion

This dilemma considers the ethical differences between environmental attorneys and other environmental professionals. This is not meant to condemn environmental attorneys, or any attorneys for that matter, but rather to better understand their professional obligations and mindset.

Regarding ethics and the law, a few years ago there was the following illuminating discussion on the San Diego County Bar Association's website:

> *Question:* An attorney represents a client in a criminal proceeding. During a meeting between attorney and client in the attorney's office, the attorney discloses to the client that a codefendant has agreed to cooperate with the prosecution and provide information that will incriminate the client. Upon learning this, the client becomes enraged, opens a brief case, displays a handgun, and tells the attorney: "I've had it with that lying scum. I know how to use this and I'm going to take care of him once and for all." The client then leaves. Based on the attorney's knowledge of the client, the attorney believes that the client intends to seriously injure or kill the codefendant. May the attorney disclose the client's intentions or otherwise warn the authorities and the codefendant of the apparent danger?
>
> *Answer:* Despite the obvious moral dilemma presented to the attorney in this scenario, California law (including Section 6068(e) of the Business and Professions Act) forbids the attorney from disclosing any information obtained in confidence from the client. No implied common law duties or rules of professional conduct promulgated in other jurisdictions apply to allow even limited disclosure.

After this dilemma was published several responses from concerned readers were received. A Los Angeles area environmental attorney wrote, "Personally, I find the opinion repugnant. I thought of a situation where a criminal client told his attorney that the was on his way to harm a policeman, district attorney, judge, or even all the way to the President. I would find it impossible to believe that the client's attorney must maintain confidentiality [in this kind of situation]." The attorney proceeded to research the topic and discovered that in 2004, the California legislature made an exemption and revised the California Business and Professions Act to allow attorneys to reveal an imminent threat of physical harm. Section 6068(e) of the Business and Professions Act, which became effective July 1, 2004, now reads:

> (1) ... to maintain inviolate the confidence, and at every peril to himself or herself to preserve the secrets, of his or her client.
>
> (2) Notwithstanding paragraph (1), an attorney may, but is not required to, reveal confidential information relating to the representation of a client to the extent that the attorney reasonably believes the disclosure is necessary to prevent a criminal act that the attorney reasonably believes is likely to result in death of, or substantial bodily harm to, an individual.

It is interesting to note that this revised code still does not allow an attorney to report of imminent harm to property or the environment. Note also that this applies to the state of California only. Other states may or may not have dealt with this issue.

Historically, the maintenance of client confidentiality has been one of the highest duties of an attorney. Society originally derived this duty, in part, from Roman law and carried it forward through common law and legislative promulgation. In California, Section 6068(e) of the Business and Professions Act provided that it is the duty of an attorney to maintain inviolate the confidence of his or her client and, at every peril to himself or herself, preserve the client's secrets. The only arguably relevant exception is the "crime-fraud" exception. This provides that there is no privilege if the services of the attorney were sought or obtained to enable or aid anyone to commit or plan to commit a crime or a fraud. Many judicial and ethics committee opinions have made it clear that the attorney's duty is to maintain all client disclosures of past crimes confidential, and that the attorney's duty also extends to protect confidential disclosures of an intent to commit future crimes and frauds, and until 2004 included even those encompassing threats of great bodily injury or death.

This is clearly an extreme case, but it helps illustrate some of the differences between the priorities of attorneys and those of environmental professionals, whose first responsibility is to protect public health.

6.9.3 Recommended Action

In this case, the answer is obvious, but the issue at hand is how to make sure that public health is protected, while minimizing any adverse impacts on you or your company by your actions. The prospective future use of the property was not specified. Since there is no specific evidence of the contamination, you must compile a strong case for estimating the degree of contamination that is likely to be present and recommend that a phase II assessment (i.e., soil sampling, etc.) be conducted at the site. In the past 30 years, there have been a number of CERCLA remediation sites with extensive data revealing the source, chemical composition, location, and toxicity of pollutants that you believe exist at the site. You must do your best to identify the specific processes that existed at the plant by locating old records and interviewing former employees, as prescribed in ASTM E 1527-00, the specification for conducting ESAs. Photos of slag piles and any other residual evidence should also be included in your report.

Assuming that this work has been done, you need to explain the results to the attorney. You must tell him that, in your expert opinion, the report reflects the best assessment of the site contamination without bore samples. You should suggest that the impact of this contamination on the future of the site should be further explored, including a health risk assessment. Explain that, depending on the future use, the amount of remediation required will vary. You have an obligation to present your best assessment of the state of contamination at the site. What is really needed is a full remedial investigation and feasibility study. The buyer must be made aware of these concerns. Because of the short time frame available, you cannot change the facts in your report, but with additional time and funding, you could at least characterize the remediation that would be required.

Consistent with the first responsibility of an environmental professional to protect public health, a full objective judgment and explanation must be presented. Anything less would be a breach of ethical values. Since you have no clear evidence of contamination at the site, and are using the results from other locations to form your opinion, it would be appropriate to qualify your opinion as to the probability of actual contamination and suggest a staged soil sampling to look for preliminary findings that would justify a more invasive investigation.

However, be sure to avoid the tendency of some investigators to become entranced with their theory of contamination. Statements can be made that are more conclusive than speculative. Therefore, if the attorney challenges specific wording as "over the top," it is not a breach of ethics to consider whether you may have been overly assertive in your conclusions without solid evidence. This is not to suggest that you water down the evidence in your report, but you may find that the use of speculative words like *potential* and *possibility* are more truthful than unqualified positive statements.

6.10 Expert Has to Balance Client Confidentiality and Medical Need of the Affected

6.10.1 Dilemma

You are an environmental professional with a private practice that specializes in indoor air quality. One day, the facility manager of GTCO, a local insurance company, calls you in to evaluate a water intrusion problem in the company's fifth-floor lunchroom. Upon examination, it appears that the drain to a sink in the lunchroom has backed up, causing water to leak into the lunchroom under the floor tile. Water is also leaking into the wall behind the sink, which separates the lunchroom from an adjacent executive office. There is no information regarding how long the problem has existed. The problem was discovered only when a bookcase in the executive office was moved, exposing mold growth on the connecting wall. Your initial investigation is scheduled after normal workday hours when GTCO's offices are unoccupied, and you are warned not to discuss your investigation or any of your findings with the company's employees. Your investigation reveals toxic mold under the lunchroom floor tile, in the ceiling tile of the floor below, and in the adjoining office wall. You submit a detailed remediation plan to the facility manager.

The remediation work is performed the following weekend, again when the company GTCO's offices are unoccupied. During the remediation, the affected area is properly isolated and ventilated, so that mold spores are not dispersed via the building's heating, ventilation, and air conditioning system. You supervise the remediation work yourself and perform a post-remediation clearance test, which shows that the air in the building has a level of mold consistent with nonproblematic buildings. Upon completion of the remediation, you are contacted by outside counsel, who directs you to maintain strict confidentiality in regard to the entire project. You are assured that this confidentiality is purely a precaution to reduce the potential for employee actions.

Six weeks later, you are contacted by an attorney representing several of the insurance company's employees, who, the attorney claims, are suffering from serious fungal infections. The attorney learned that you had been engaged to test the building for mold and asked to see the results of your investigation. You explain that it is your practice to provide no comment when asked by an attorney about any case involving a client. The attorney urges you to reconsider, saying that some of the employees are seriously ill and that their doctors need to know what they have been exposed to in order to treat them effectively. How should you proceed?

6.10.2 Discussion

The law is clear in this case. You had signed a contract with GTCO and agreed to maintain strict confidentiality. Your initial response to the employees' attorney was appropriate. You did not acknowledge anything, nor did you reflect on any specific case. That was correct and demonstrated a strong professional approach to the situation, but now the attorney tells you that there are cases of serious illness (assume that this means potentially fatal). While it is the primary responsibility of an environmental professional to protect the public, that responsibility, like any other pillars of character, must be properly weighed and measured, and should not be used as an excuse to act inappropriately. You do not know the caller, and this is the first you have heard about these cases of serious illness. The illness claims could be a shameful ruse to gain your sympathy. It is important not to act spontaneously in this situation. Caution is paramount. Take a moment to think the situation through to avoid reacting according to instinct.

6.10.3 Recommended Action

As a consultant in this case, you hold a position of trust with your client, GTCO. Your pledge of confidentiality must be respected to the fullest extent possible. The outside counsel working on behalf of GTCO should be consulted as soon as possible. Report to the GTCO counsel the details of the call you received from the employee's attorney. Inquire whether any of the employees have reported their illness to the company, and if so, what the company is doing to resolve the issue. You do not know the circumstances under which the employees' attorney was aware that you were the environmental consultant in the case. It is appropriate, therefore, for you to inquire whether your test data—that describe the mold species to which the employees were exposed at the time of the test—have been provided to a treating physician or other third party.

Ask GTCO's counsel how you should proceed in providing expert assistance in the case—in particular, how you should respond if you are contacted again by the employees' attorney. You must refrain from offering medical advice. However, you may offer an interpretation of the test data. Since mold spores are ubiquitous, you should explain that the concentration of spores indoors must be compared to the concentration outdoors to evaluate whether mold is growing inside the building, that mold concentrations can change frequently, and that your test results serve only as a snapshot in time. You should explain the probable cause of the mold growth and the steps that were taken to limit the migration of mold spores during the remediation work. In addition, you should offer any other information that the counsel may request in regard to exposure and remediation. Depending on the degree of tension between GTCO and its affected employees, you may or may not be given details regarding employees' illness. It would be sufficient to just know that it had happened. Assuming that you are satisfied that the firm has handled the situation appropriately, you will have fulfilled your ethical obligation.

In this case, your professional and ethical obligations to protect public health were fulfilled by using your expertise to determine the cause of the mold, the degree of exposure, the remediation protocol, and the precautions taken to prevent further migration of the mold spores in the building. The employees who were affected by the mold exposure prior to the remediation work have consulted medical experts and are being treated accordingly. The details of this treatment are not of concern to you, and neither is the responsibility of the company to compensate the employees for their illness. However, if, in your opinion, GTCO's facility manager has not dealt with the situation fairly, then it is your ethical responsibility to report your concerns to GTCO's higher management. If you were part of a consulting firm, you should request your company's attorney to write a letter to an appropriate member of GTCO's management. Since you are an independent consultant, and if you do not receive a satisfactory indication from GTCO that it has acted fairly in this matter, writing a registered letter to that same top management member should be considered. Beware, however, that if you decide to provide details of your investigation to the employees' attorney without permission from the insurance company, you could open yourself and your consulting firm up to a possible lawsuit for breach of contract.

The insurance company has your report and the employees' attorney can subpoena it.

6.11 Firm Faces Welfare Extortion via Disingenuous Protest from Interest Group

6.11.1 Dilemma

You are the environment, health, and safety (EH&S) director of Wouldbee Petroleum, Inc. (WPI), a refinery start-up located near a major U.S. metropolitan area. WPI is newly incorporated and recently acquired ownership of a small refinery that was shut down by its former owners after several years of struggling to operate in a loss mode. Your firm purchased the plant, hired expert management to assess the plant's existing operations and the future market for refined products, and concluded that with an initial investment of several hundred million dollars, the plant could be highly profitable in a few years due to increasing demand for refined products. The primary hurdle to overcome is the refinery's bad reputation for unhealthful releases caused by the former owner's failure to perform proper maintenance (due to the unstable financial situation). As EH&S director, you negotiate with the EPA, state and local agencies, and city administrators to obtain the necessary air permits, including the approval of an environmental impact statement and conditional use permit.

The refinery would generally be seen as an employment boon for the local community, providing more than 1,000 new jobs for nearby residents, many of whom worked for the refinery in the past. Local city officials are also supportive. Despite this support, however, a group of attorneys representing local plaintiffs, who call themselves Local Environment Resistance (LER), maintain serious opposition to the reopening of the refinery, filing restraining orders and claims of regulatory violations. The group has tried unsuccessfully to sue WPI and the regulatory agencies several times. Each time LER lost a court decision, it refiled another suit within a few days.

One day, LER's lead attorney invites you to lunch in a remote area of the city to discuss the situation, where he suggests that if WPI were to contribute significantly to LER's welfare fund and provide free health care for its members, LER would withdraw its opposition to the refinery. You are warned that "this meeting never happened" and that any initiation must come from WPI. You take this offer to WPI's senior management and a meeting is convened. Your company is faced with extortion on one hand and a questionable investment of many millions on the other. How do you proceed?

6.11.2 Discussion

This dilemma requires serious thought and discussion. There is no easy or obvious answer. On one hand, WPI has already invested millions of dollars in the project, having employed a technical and support staff of 100 people for over a year, performed business analyses, and proceeded with process design changes. Government agencies have been kept well informed throughout the process and are supportive. It is financially critical to the sponsors to proceed with the project. On the other hand, it is unethical, if not illegal, to pay off LER, even if the offer could be dismissed as an ordinary out-of-court settlement. A colleague expresses concern that to pay off LER could be the beginning of continual pressure from the group for additional payment. The problem is that there is no evidence that LER has solicited this payoff. If the payoff is exposed, WPI risks being charged with improper conduct by its investors. If the offer is ignored, the legal costs—already over a million dollars—will likely continue to mount. And if LER should succeed in obtaining a restraining order in the future, the effect on the project would be devastating. The final consideration is obvious: What if WPI decides to withdraw from the project? Although further investigation is required to confirm the details, it is estimated by WPI's controller that the value of the existing process equipment and the real estate value after the necessary "Brownfields" remediation would allow WPI to come close to breaking even.

6.11.3 Recommended Action

WPI's choices boil down to the following:

1. Reject the offer and continue the project, but respond to maintain favorable judgments in the courts until LER is finally forced to desist by the courts
2. Reject the offer and discontinue the project, selling off the equipment, remediating the property, and selling the real estate or developing it for an alternative use
3. Accept the offer and use whatever options are available to ensure that LER cannot renege on its commitment, and take precautions to avoid the presumption of deceit

Which option would you choose? Do you see another alternative?

Accepting LER's offer and negotiating a monetary settlement to make the group go away quietly, the third option, violates the ethical values of trustworthiness, respect, and responsibility. WPI would constantly have to worry about being discovered. If LER would agree to a public declaration of its withdrawal in return for certain benefits, then the situation might be more acceptable. The declaration would help explain that this is in the best economic and employment interests of the local community. Considering that the regulatory agencies believe that the proposed design changes will eliminate objectionable environment impacts, the refinery would be considered an overall benefit to the town from a business standpoint.

If LER is not willing to accept an open and public arrangement at an affordable price, then its objections should be made public and the support of local merchants and citizens groups should be solicited to counter LER's opposition. This is concurrent with the first option mentioned above. The primary disadvantage of this approach is that it might discourage outside investors. The fact that there is ongoing litigation would make it difficult, if not impossible, to issue stocks or bonds, for example. Alternatively, if a large investment entity is willing to finance the project because it sees the potential for profit, then the litigation becomes just another cost of doing business. Given WPI's past successes in court, the investor could probably assume that LER's charges would eventually be dismissed or invalidated.

In the end, if no investor materializes, and WPI's efforts to sell investment securities are unsuccessful, or if the poor environmental performance of the previous owner has caused the neighbors to lose trust in any refinery operator, then it would be an act of civic virtue to discontinue the project, particularly if the money invested so far can be recovered from sale of the equipment and real estate.

While it might be tempting to pay off adversaries to new projects with potentially correctable environmental consequences, the practice is unethical and risky. Therefore, assuming that this refinery would be highly profitable and beneficial to the local community, the ethical practice is to make the offer known publicly. If going forward with the project is the right thing to do for all stakeholders, then the courts should ensure that legal justice is done. If, on the other hand, the project has a high-risk potential in the minds of the local residents, then the "right thing" is to discontinue the project and settle for the intrinsic value of the site's equipment and real estate.

6.12 Plant Manager Installs Emission Control Equipment without Proper Permit

6.12.1 Dilemma

You are director of environmental affairs for ABC, Inc., a major manufacturing corporation with 27 plants in the United States and 5 outside the United States. During your tenure, you have instilled a strong policy of compliance throughout the company, including certification under ISO 14001 and its philosophy of continual improvement. Your ABC corporate staff makes routine visits to the plants to audit their individual programs for compliance. During a routine audit of a large plant in Nebraska, your staff reports that in the past two years, the plant has installed two baghouses and a sulfur dioxide (SO_2) scrubber. All three control systems have been source tested, are continuously monitored, and are shown to comply with their intended performance. However, the auditors discover that none of the new systems have the required state permits or are included under the plant's federal Title V permit, which is also a requirement. During your own visits to the plant, you were so impressed with the plant management's initiative and the technical performance details of the equipment that you had no reason to suspect this type of permitting discrepancy. You even reported the plant's achievements to your company's board of directors, which issued a commendation to the plant manager. You wonder how this embarrassing situation could have occurred and what should be done about it.

6.12.2 Discussion

This dilemma is based on a frequent real-world situation that deals with an overzealous environmental manager. You contact ABC's plant environmental manager for an explanation of the oversight. His response was as follows:

> I realized that, based on my calculations, our prior controls—dry cyclones—were not doing the job of removing particulate matter and SO_2 to meet regulations. As a result, we were not in compliance. The agency was apparently unaware of this. I was also concerned about mercury emissions. In the ISO spirit of continual improvement, I convinced the plant manager to install the necessary new controls. Since permitting would draw the agency's attention to this situation, we avoided the problem by not submitting a permit application. We have a good relationship with the agency inspectors, keep our monitoring data up to date, and seldom, if ever, exceed our limits. We are the largest employer in town and have great public relations with the local community.

One might classify this dilemma under the category of "no good deed goes unpunished." While the equipment is in technical compliance and the protection of environment and public health is preserved, it is clearly illegal to operate the control equipment without a permit.

6.12.3 Recommended Action

The recommended approach to remedy this oversight is self-reporting. Depending on the plant's relationship with the local agency officials, either a plant official or corporate attorney should contact the appropriate agencies immediately. It is important that the situation be presented in a positive light, explaining the benefits of the new control equipment. If the plant has an otherwise excellent record of compliance, as must be the case under ISO 14001, this should be emphasized. To encourage self-reporting, most agencies significantly reduce fines for these types of violations. If indeed the plant has an outstanding environmental record, the agency should be willing to work with the company, mitigating the penalty and sparing the embarrassment that would ensue.

The plant personnel, in spite of their good intentions, must be admonished for this violation. You should hold a meeting with all of the personnel involved to determine who specifically made the decision to avoid permitting. The environmental manager surely knew about the discrepancy, or should have known better. It was his responsibility to ensure that the proper regulatory procedures were followed. You should have been informed so that you could have persuaded the plant manager to follow the necessary permitting procedures, or if necessary, report the discrepancy to top management. Of course, the proper course of action would have been to go to the agency in the beginning with a proposal to install the control devices, explaining the need to expedite the permit approval. All of the staff involved, including the plant's environmental manager, should be criticized, being sure to consider their previous performance and their intent in this case.

6.13 Consulting Firm Bills Client at a Higher Senior Rate for Junior-Level Work

6.13.1 Dilemma

You are a senior environmental consultant at Clean Environment, Inc. (CEI), a firm with many offices and employees. You have a number of special technical skills and licenses that make you especially valuable to your firm. CEI's management is pressing the firm's supervisors to maintain a high billability percentage. The clients, for whom CEI has worked extensively for several years, are experiencing difficulties in the current economic climate and have discontinued using outside consultants. Your company has assigned other billable work for you to handle. However, you realize that the work is simple and routine and could be readily handled by a junior-level project assistant. You turn in a time card and the client is charged for your time at your senior consultant rate. You are concerned that the client is being unfairly charged for the work that can be performed by a junior person. You speak to your supervisor, Jan, about your concern in a private conversation. Jan's response is that she is just protecting you from a layoff and the project's budget can afford you because she discovered some existing modeling work that could be applied to that project. This saved significant new modeling labor that had been included in the original proposal budget. Although this is a cost-reimbursable contract,

Jan feels she can use the budgeted funds to support her staff as long as the total estimated amount is not exceeded. Is Jan's action ethical?

6.13.2 Discussion

In questioning the ethics of this practice you suggest that the client be charged a fair rate by lowering the overhead burden and the profit. Jan reasons that even if a junior-level project assistant were used, that person would have to catch up and train while you already know the routine. The cost to the client for this extra training would be a push. You argue that the client should not be charged for training the junior person since the client was told that the work would be done by experienced personnel. The conversation ends when Jan says that she is under management pressure to maintain billability and she is doing her job, which she has to get back to in order to maintain her own billability.

This is a dilemma that has been faced by many environmental consultants over the years and is likely to be a prevailing issue right now with the business downturn that the world economy has been experiencing. In this case, the ethical values at risk are trustworthiness and responsibility. Trustworthiness implies honesty. Falsely charging a client is dishonest. The seventh pledge in the A&WMA code of ethics states: "exercise honesty, objectivity, and diligence in the performance of all my professional duties and responsibilities." Responsibility requires that you recognize this situation and take steps to correct it.

6.13.3 Recommended Action

The recommended action to be taken is to arrange a meeting with Jan and suggest that she tell the client that she has chosen to use you to perform the assigned tasks because of your experience and the opportunity to get you acquainted with the project so that you can perform some other tasks. She should explain the added value of using you to do this project instead of a junior person. Often in a project there are tasks that need to be performed that were not contemplated at the outset, and therefore are not in the budget per se. Jan should show the client that she was able to save project money by using the existing modeling results. As a result, she can now take advantage of your special qualifications and project experience to have you perform the unanticipated task because you now have a good grasp of the program. She should explain that the client would receive all of that extra work without overrunning its budget. Suggest that by being honest in this manner she will gain the client's respect, and thereby obtain future business. Listen carefully to her response and try to make this her idea. Encourage her to apply whatever spin she feels would help sell the plan.

If Jan is not willing to go along with your idea, suggest that she and you go to her manager to review the suggestion. When working for a client, be alert for problems it is experiencing that you could solve for it. When a sum of money has been budgeted for a project and you recognize an underrun, be prepared to offer to use the surplus money for some constructive purpose rather than let it be returned. This is honest and helps both parties.

6.14 Firm Offers Country Club Membership for Procurement Officer

6.14.1 Dilemma

You are the manager of a 50-person office for Environmental Consulting Unlimited (ECU), a regional environmental consulting firm. A former client, Widgets-R-Us (WRU), is planning to build a new manufacturing plant in your area. In the past, WRU has engaged your company for work outside your district. ECU's corporate marketing staff has discovered WRU's intentions to build the new plant and is aware that several firms will compete for the permitting and emission control design and fabrication contract. You have responded to WRU's request for proposals. WRU's plant manager-to-be, Brian, who has been assigned as project manager for the design, permitting, and construction of the new plant, comes to your town to scout the area. You meet him and begin a dialogue. You introduce Brian to the local agencies and brief him on housing opportunities, civic organizations, and local recreation. You are aware that some of your competitors are also giving Brian special attention. Your corporate marketing manager informs you that two of the competitors have also been wooing WRU and will give you a serious run for the money on this bid.

During a strategy session with corporate marketing staff, you brainstorm the steps that can be taken to give ECU an edge in this competition. You know that Brian is a keen golfer and has seen the country club where you are a member. You suggest that Brian be offered a complimentary membership with the idea that there will be other client management staff moving to the area that might also join if Brian is a member. You suggest that ECU call the country club manager indicating that ECU will make a donation to help cover the cost of Brian's free membership. ECU's marketing staff authorizes this arrangement. After the proposals are submitted, the three vying companies are each invited to interview with WRU before a final selection is made. During the interview, you mention to Brian that the town is looking forward to WRU's arrival and inform him of the planned complimentary membership. Brian's reaction is, "That would be nice." You say, "I look forward to playing golf with you." Is this action ethically acceptable?

6.14.2 Discussion

First of all, understand that ECU has not requested special consideration. The golf club membership will be offered as a promotion regardless of which company is selected by WRU. Consequently, should this offer of free membership to your country club be considered a bribe? A bribe has been defined as something that is given to another person in a position of authority or trust to induce that person to behave in a manner inconsistent with such trust. Does taking these steps to make Brian feel welcome seem sufficient cause for Brian to violate his position of trust at WRU?

On the other hand, since you were the person who informed Brian about the arrangement, it is likely that ECU will receive some benefit. If the offer was just to promote a smooth working relationship in the future, why was the offer not withheld until after the contract was awarded? Clearly, there would be more opportunities for consulting work in the future. A close relationship with Brian would be advantageous to ECU in terms of obtaining future work.

Is this offer any different from a promotional discount at a furniture store, for example? Yes, it is. The country club membership is not being offered to WRU as a customer company. It is being offered to one WRU employee with no clear benefit to the company. Also, it is being offered at the request of ECU, one of the competitor companies. This is not the same as using a discount as an incentive to encourage a customer to buy a product he or she may not have otherwise considered. So, is ECU's action in this competitive action ethical?

6.14.3 Recommended Action

In this hypothetical case ECU's action has been taken and must now be judged as to its ethical acceptability. The question of ethics should have been addressed up front, during the strategy session when you agreed to arrange for the complimentary membership. Brian is in a position of trust with WRU. You are offering something that is only of benefit to Brian, but which your marketing staff hopes will be influential in the selection process. It was a clever move to not predicate the country club membership on the selection of your firm. That would be clearly unethical. In that case, if your firm was to be selected and the membership deal goes through, the competitors could find out and protest the contract award. If WRU top management learns about it, it could be embarrassing for Brian.

From an ethics standpoint, does this special arrangement for the incoming plant manager—even though it was not directly predicated on any commitment to award the contract to ECU—create an unfair advantage for your firm? It was clearly the intent in the strategy session—where the idea for the complimentary membership surfaced—to improve ECU's competitive edge. By not tying this special privilege to the result of the competition, it would be hard for the competitors to protest an award to ECU. But there should be no doubt that Brian would recognize who initiated the idea and made the arrangements for his free membership. For that reason, it is unethical for an ECU representative to privately arrange this type of client benefit.

6.15 Employee Notices Dishonest Use of Company Resources by a Colleague

6.15.1 Dilemma

You are a graduate environmental engineer who recently joined Industrial Products, Inc. (IPI), a large manufacturing firm with plants across the United States. You share an office with another employee, Bill, with whom you also share assignments from a common supervisor. You soon become aware that Bill takes advantage of opportunities to cheat the company. During your first field assignment together at an out-of-town plant, you learned that Bill was claiming mileage even though he rode with you. He also boasts that he has inflated his motel room bill by obtaining a blank receipt and filling in a higher price. In addition, you are aware that he spends hours at a time playing computer games or just idly surfing the Internet while on company time. On another field assignment together, he mentions that he pads his time card and charges his partying expenses to his expense report. When you question him about these practices, which also include taking home office supplies, his response is that no one cares and that the company will never miss them. When you suggest that what he is doing is stealing, his reply is, "Come on, grow up, everybody does it."

Bill is affable and seems to have a joke for every occasion. In meetings with your supervisor, Jesse, Bill speaks up frequently and appears bright and knowledgeable. It is obvious to you that Jesse likes Bill. If Jesse has ever questioned Bill about his time card or expense account charging practices, you are not aware of it.

You begin to meet a colleague, Roger, from another department of the company to play handball twice a week. One day, the subject of your office mate comes up in conversation. Roger says that he and Bill joined the company at the same time and had briefly discussed their previous employment. You learn that Bill had moved from job to job nearly annually and that he has been at your company for almost two years. You feel close enough to Roger to have a confidential discussion in which you discuss Bill's cheating. Roger is in manufacturing and works on a strictly 40-hour-week basis. He says that he has heard that the field employees pad their expenses, but he never gave it a second thought. "Far be it from me to blow the whistle on those guys. Whistleblowers become pariahs. If I were you, I'd keep my nose out of it and let the system catch him. After all, Jesse signs off on his expense account. Since Jesse doesn't catch him, it could make your boss look bad if you report it to a higher authority."

6.15.2 Discussion

When posing this dilemma to a number of friends and colleagues of the authors, the reactions have ranged from "don't do anything, it isn't about you, and the risk is too great to take it on" to "you have to tell the company, because that person is cancer that can eventually affect the whole staff." The authors request you, the reader, to consider your own response before you read on.

People usually do not deliberately set out to cheat the company. But opportunities surface that test a person's honesty. If cheating on an expense account becomes routinely acceptable, then what if a client suggests to an engineer that he fudge test data to pass an agency's requirements? You can see where this might lead.

On the other hand, if you report Bill's cheating practices to your supervisor, it is likely to get out that you told the boss. If disciplinary action is taken on Bill, he and other colleagues might look upon you as a tattletale, and things at the office could get ugly. Instead of being about Bill, the issue now becomes about you. This level of cheating is hardly worthy of branding you a whistleblower. Nevertheless, you do have some responsibility to your company and to your clients to see that they are not being cheated.

As environmental professionals, your main priorities are to protect public health, be loyal to your company or clients, and be considerate of your loved ones. The values involved in this case are trustworthiness, respect, responsibility, fairness, and caring. Trustworthiness involves being loyal, honest, and reliable, and having the courage to do the right thing. Responsibility involves doing what you are supposed to do and being accountable for your choices. Respect is following the golden rule: treat others as you would have them treat you. If you were Jesse, wouldn't you want to know what is going on? Fairness is playing by the rules, not taking advantage of others, and not blaming others carelessly. That is why one must not be careless in placing blame. Caring involves being kind and compassionate, forgiving others, and helping others in need. Is Bill a person you could help get back on track and recognize that cheating at any level is wrong?

6.15.3 Recommended Action

Before doing anything else, find someone you trust to discuss the issue in confidence. It is important to get feedback before making a decision on what action to take. The person you confide in may not agree with your decision, and you can explore that person's rationale to see whether it agrees or conflicts with your own.

The person you confide in might suggest that you first go to the appropriate authority (assuming it's not your boss, Jesse) and request a private office or a rearrangement of office mates. If you must provide a reason for moving offices, indicate that you have a hard time concentrating on your assignments with Bill in the room, but avoid mentioning Bill's cheating practices.

Now, here's the important action: to handle a colleague's unethical behavior without coming off as a tattletale. Most of the other dilemmas can be handled by the ethical person without involving others directly. At the next meeting of your group or in a private meeting with your boss, Jesse, suggest that your group begin ethics training. You can start small, for example, with brown-bag seminars once a month. Explain to Jesse that there are many publications and professional code of ethics that can be provided to the group as subjects for discussion in the seminars. If this approach works, then after a few sessions, introduce this dilemma and discuss it with the whole group. It may help to give Bill a new attitude. Bill will likely never recognize himself in this article.

6.16 Hazardous Waste Is Accidentally Sent to a Municipal Landfill

6.16.1 Dilemma

You are the environmental manager of Specialty Chemicals, Inc. (SCI). Your specialty chemicals are mostly of the hazardous type, and any spent chemicals are properly packaged, sealed, and sent to a licensed facility for disposal. On the day before your vacation started you properly labeled and sealed a drum of hazardous waste and stored it in the "Hazardous Waste Accumulation" area. You also made arrangements with a certified transportation and disposal company, Hazmat Transit Co. (HTC), to pick it up while you were away. However, when you return, you inquire at HTC about the disposal of the drum and learn that somehow a mistake was made and the drum was delivered to the municipal landfill. What do you do?

6.16.2 Discussion

In factories where hazardous substances are handled on a daily basis and there are formal procedures and training made to insure worker safety, the result is that often the plant personnel take a relaxed attitude about dealing with those toxic chemicals and waste. If SCI's plant manager, Dave, knows of the problem but has taken an attitude that "accidents happen and it is unlikely that anyone at the landfill will notice the barrel labeling and correct the results of the mistake," you cannot let this stand. If you cannot convince

Dave to report the incident and take remedial action, you should escalate the issue to his or her supervisor or possibly to the company's attorney.

Finally, reviewing the six pillars of character, the primary one in this case is responsibility—to protect public health. Emphasize the seeming relaxed attitude that has become apparent to you and that such a casual attitude could possibly lead to a serious accident and possibly a loss of life. Putting this attitude in a formal memo to all hands should stimulate further concern and comments. Explaining your concern for public health in a nonhostile manner to both the plant manager and the company attorney should motivate them to accept your approach. Taking the necessary action to prevent a recurrence should also gain their backing. Trustworthiness is also involved in maintaining an honest appraisal of the situation. Hopefully your straightforward approach to handling this problem will be a demonstration of your character and gain you the respect of your management. Justice and fairness are factors when investigating the cause of the misshipment and what discipline you will recommend for the person(s) who allowed this unfortunate event to occur. Caring and civic virtue and citizenship are pillars that are implied in the actions taken. The actions taken to protect the public and report the incident to the local agency demonstrate these two pillars.

6.16.3 Recommended Action

First you must notify your company management, Dave, and the other department heads. Then you must notify the municipal waste landfill and the hauler to try to locate the area where your waste was offloaded. You must describe the containers and contents in order for the waste to be located. You should provide sniffer instruments if appropriate to detect any raw waste that might have been unpackaged. In short, you must do whatever is possible to prevent the hazardous waste from being released where it could affect public health. Request a formal response from the landfill manager as to whether or not the material was located and what has been the disposition. In parallel with this effort you must notify the local hazardous waste management agency to report the incident and accept further direction from that agency. This is the urgent part of your response.

If even after the hazardous waste misshipment is confirmed beyond any doubt in your mind, Dave and the senior staff are unwilling to do the right thing and report it, then you should perform a thorough investigation to learn the cause of the incident and prepare a formal report of the incident with all available supporting documentation. Then you should submit copies to Dave and his immediate superior(s), the company's attorney, and other responsible parties, such as the transportation personnel who ordered the pickup. Assuming that your management responds favorably to your corrective measures (as presented above), you must determine who failed to do their job of properly carrying out the hazardous waste pickup and recommend disciplinary action for the person(s) responsible. This could be a lesson to your fellow employees on the importance of ensuring proper disposal of hazardous waste.

6.17 Risk Communication—Numbers but No Panic

6.17.1 Dilemma

You are the public relations officer (PIO) of an electric utility Commonwealth Electric (CE). In 2002 your plant discontinued the use of chromium compounds in its cooling tower after the practice was banned. The state has an air toxic program that requires (1) reporting of air toxic emissions, (2) conducting health risk assessments if the air toxic emissions are above a certain threshold, and (3) notifying neighbors if their health risk exceeds a certain level. Until 2002, chromium emissions from the cooling towers were a major contributor to the health risk from this plant. Because of the slow execution of the program, CE is now required to notify approximately 10,000 residents near the power plant that in the year 2001, they were exposed to a cancer risk of over ten cases in a one million population with a maximum exposed individual risk of 187 cancer cases in a million. The air agency said that it has no recourse under the law. It has to be reported. It allows you to indicate the fact that the chromium is no longer used and that the risk has been significantly reduced in your risk communication letter. CE's environmental department indicates that, without the chromium, only 1,000 homes would be above the ten in a million notification threshold and that the maximum exposed individual risk would be 37 in a million. You are also told that recent emissions tests indicate that future health risk estimates may exempt CE's plant from any further notification. How would you write the letter? Consider this question: What is greater, the actual risk of cancer or the fear of cancer?

6.17.2 Discussion

This dilemma is more a function of regulatory transition and formal protocol than it is of protecting public health. If the cancer rate in this region, as determined by the local health agency, has not been beyond acceptable limits, then stirring up public concern can have an adverse effect. Since the exposure level has been reduced by the discontinued use of chromium, reporting the former values cannot have a useful benefit. Occasionally the local agency staff will quote the current regulation without informing the inquirer as to the limits of applicability. Alarming the residents when there has been no apparent excessive health risk is just as wrong as lying. On rare occasions, lying can be the right thing, e.g., the case of killers chasing a potential victim and asking you which way the victim ran.

While it may be off the main subject of this dilemma, you should take this matter to a corporate environmental advocate and attempt to obtain a variance from EPA citing the prospects of creating anxiety among CE's ratepayers. If this is unsuccessful, take your request to the local congressional office to see if the staff there can be convinced to speak to the EPA for a deviation to this requirement.

All that said, if you, as PIO, had to write the letter to the 10,000 homes, the recommended phrasing would be as presented below. If you have preferred wording, discuss it with you colleagues.

6.17.3 Recommended Action

The letter should read as follows:

> This letter is being sent to you to comply with a state-mandated notification. Since 1990, the federal Environmental Protection Agency (EPA) has been investigating the health impacts of a multitude of air pollutants that have a potential adverse health effect on exposed individuals. One such toxic substance is chromium. Chromium was used in our cooling towers in the past to improve their performance. However, once CE learned of the EPA findings that chromium was toxic, CE's use was discontinued (in 2002). Since then the state agency has developed an air toxic program that requires the notification of neighbors to CE's plant that have been subjected to a cancer risk of over ten cases in a one million population. Currently the official estimated emissions from our power plant in your area would exceed ten in a million in approximately 1,000 homes. However, more recent test results indicate that even this estimate is too high, and soon there will be no increased cancer risk from our facility. At this point in time, the greatest risk in this area officially is 37 in a million. Because of the timing in having the state's regulation issued we are required to notify the residents who were exposed ten years ago when the company still used chromium. At that time, the exposure to the emissions from our power plant in your area that exceeded ten in a million cancer risk was approximately 10,000 homes, and the greatest risk in the area was 187 in a million.
>
> Please understand that the risk assessments in this case are extremely conservative. For instance, they are based on a person remaining in the home 24 hours per day for a period of 30 years. Please take care to fully understand this notification, and if you have any questions, contact us at (phone number and email address).

6.18 Agency Inspector Ponders—To Cause a Plant Closure or Not

6.18.1 Dilemma

You are an agency enforcement inspector with 20 years of fieldwork experience in a small one-industry town. You are the lead person in this agency branch office of five persons. You find a significant infraction at the Murphy Magic (MM) plant that is the town's principal employer. You know many of these workers from your church and club membership, and you are aware that MM is struggling financially. The town's population is approximately 5,000. For many years the waste chemicals have been deposited in a landfill operated within the plant property but not sanctioned by your agency. You might have never learned of this illegal landfill except for a comment you overheard in a conversation at your club. You learn that the fine for this infraction would surely force the plant to close, leaving a residual contamination for the government to clean up, and would result in the layoff of 800 workers.

The plant owner, Conrad Peak, meets with you and requests a week to deal with this situation before you report the infraction to the agency. At the meeting a week later Conrad had received information as to the cost of the violation and the necessary cleanup effort. He offers to work on a cleanup plan if you do not report this infraction. He has arranged for financing and believes that they will have a significant order for MM's product line in the near future.

You are a family man with three children in high school and college. Your agency pay works well in this small town and you have saved some toward the children's education. If the plant were to close, you feel that you would be kept around to oversee the shutdown, then transferred to the larger main office in the nearby city.

You want to do the right thing. What action do you take?

6.18.2 Discussion

This dilemma deals with a longtime civil servant in a small town outside of a large city who faces a decision on how to deal with an enforcement action that could have a severe impact on the neighbors.

This is clearly a dilemma that you cannot handle alone. Your commitment is to the government, and to fail to report could cause your immediate dismissal. On the other hand, the people in this town are your friends and you care about them. You had never thought about the existence of a landfill. Toxic waste had never been an issue until you just came upon it by accident. Only the five of you in the local office know of the problem, and you have their commitment to not reveal any information. You have not told your family, although they wonder why you cannot sleep at night.

6.18.3 Recommended Action

As a trustworthy employee of your agency, it is your responsibility to report your findings to the head of your agency, Mary Hale. You discuss with her an approach that would protect public health in the future and possibly keep the plant from closing. You present your knowledge of the town's financial and residential situation. There are many factors to be considered before a decision can be made. Indeed, the plant is the principal employer in the town; the issue of corrective action must be discussed between Mary and Conrad Peak. If the potential discharge from the plant's landfill is large enough, there may be federal aspects to this violation and the EPA may need to be consulted. It is doubtful that anyone will want to cause the plant to close if a reasonable alternative can be determined. But the obligation to protect public health is a first priority. The values of justice and fairness and caring are of ultimate importance in this situation, as well as civic virtue and citizenship.

It will be necessary to assess the extent of the damage to public health caused by the violation and determine the cost impact of any remediation, if possible. If equipment must be modified or replaced to comply with the regulations in the future, the costs of these actions must be assessed and a means to finance them determined. You should use your expertise to help define the issues and the possible corrective actions to be taken. You should also determine the cause of the violation and who specifically was at fault. Was it a matter of lack of awareness of the factory staff and management? Or was it the agency's failure to notify MM of the Superfund regulations? Had you been adequately trained to look for such a violation? Why it was not corrected? It will also be necessary to set up a test program to identify the toxic chemicals in the landfill and determine whether groundwater was affected. All of this information will be supplied to the Hearing board that will undertake the judicial action to determine the disciplinary action.

Whether you had cause to know about the landfill before this recent discovery, you are responsible for allowing it in your jurisdiction. You must perform the tasks outlined above, but also do everything in your power to gain the support of the state and federal legal experts to help MM though this mess. You should work with Conrad Peak to meet with the local congressional office and try to work on mitigation of the fines and possibly gain the support of the federal Superfund program, as necessary, to assist with the landfill cleanup and suppress any runoff that may threaten the water supply.

Finally, you should spend any free time interviewing the plant personnel that you know to learn how this serious event was never reported and why you never detected it.

6.19 TV Reporter Pressured to "Get the Dirt" on Local Industry

6.19.1 Dilemma

Bill is a local TV news reporter assigned to cover a 1,000-gallon spill of no. 2 fuel oil by P-Chem, a local chemical processor. Bill arrives with his camera crew and is welcomed on-site by P-Chem's plant manager. The spill is being cleaned up and the incident seems to be under control. There are, however, some dead fish, drums of oily residue, and oil-soaked pigeons. Bill's supervisor, Greta, calls him on the TV van phone as he is ready to leave and asks how things are going. He tells Greta that there is really nothing to report. She explodes, ordering him not to leave until he gets some "good dirt" on the company for the 6 o'clock news, and further states: "That company has a bad reputation as far as I am concerned. You'd better not let them get away with anything." How should Bill proceed?

6.19.2 Discussion

This dilemma deals with a TV reporter who has to get the sensational story regardless of the facts in the case. News broadcasting has become more of a creative art rather than an accounting of world and local events. Each network seems to take sides in political issues. The press used to say, "All the news fit to print." Now it is more like, "All the news we wish to print"—selective reporting. As this book was in preparation, the communication media giant Rupert Murdoch was being investigated because of illegal invasion of privacy by penetrating private communications on the wireless networks. The story of this oil spill is an example of the attitude of the media mogul's need to sensationalize issues to "create" news.

Another example of how the media has affected environmental issues is in global warming, whether it is anthropogenic or not. From an ethics point of view, it is difficult to assess which of the arguments, pro or con, is honest. Each side presents its selective truth, the supporting data, but often does not present an honest assessment giving both sides of the argument. Every heat wave or windstorm creates new opportunities to bring in expert opinions to support the preconclusions of the particular network. This dilemma brings up the issue of the responsibility of an environmental reporter to present an honest account of an environmental event.

In deciding how the reporter in this case should handle his boss's admonition to get some good dirt on P-Chem, let's look at the pillars of character. From the trustworthiness aspect the reporter should have the integrity to present an honest account of the oil spill. It is not unreasonable to pursue further interviews of P-Chem's personnel in order to learn more details on how this spill took place. The objective should be to have the integrity to report the mistakes that were made so that the public can be educated as to what is needed to avoid the problem in the future. The reporter has to have loyalty to his employer, the newspaper, but there are limits to this. Loyalty is a reciprocal concept, and one has no right to ask another to sacrifice their ethical principles in the name of a special relationship. In fact, Greta forfeits her claim to loyalty when so high a price is put on their relationship. The ethical value of loyalty does not justify violation of other ethical values, such as integrity, fairness, and honesty.

Respect is another value that should be followed. The reporter should respect his boss but also respect the P-Chem people, but "getting some good dirt on the company" is not respectful. Responsibility, justice and fairness, and caring are other values to be considered in deciding how the reporter should proceed.

6.19.3 Recommended Action

Bill must reconsider his decision to depart the scene. Greta may want a story that will get her news service recognition, or she may believe that the P-Chem truly has been careless in its past environmental compliance. This spill should have been reported, and the regulatory agencies involved should be on the scene. Bill needs to get to individuals from those agencies and determine the extent of the damage and the threat to public health. If the spill was properly reported and remedial action is underway to the satisfaction of the agency, then it will be necessary to find out how the spill occurred and if this accident could have been prevented. Furthermore, he should interview the P-Chem staff and management to learn what corrective action they plan, or at least when this plan would be available, and what disciplinary action will be taken. From the agency inspectors on-site he should inquire about previous violations. How many have there been, and are they on the agency's watch list? When all of this information has been assembled, a story for the 6 o'clock news can be told as accurately and honestly as possible. Depending on his findings in this regard, he may need to pursue an agency search for historical performance of this company. But in any case, he should not invent illegitimate behavior just to create interesting press.

6.20 New Data Blows Industry's Agreement with EPA

6.20.1 Dilemma

George is the research director for Large Manufacturers Association (LMA), a prestigious trade association in Washington, D.C. LMA develops industry standards for equipment dimensions and performance and often acts as a lobbyist for agencies when those agencies prepare regulations affecting LMA's industry members. In the event that an existing regulation causes undue cost or risk to some of those association members, the association will fund research to develop and promote changes to those troubling regulations. You have just completed such a three-year project to revise the drinking water standard for the pollutant a-benzyl carcinogen (ABC). Based on data you submitted to the EPA, the agency was persuaded to relax the standard for your industry. The new standard means that 35 percent of your facilities will not have to shut down, preserving jobs and reducing the treatment cost of the municipal water districts. During the comment period for the newly revised standard, a colleague in Holland asks George to peer-review an international study on ABC. The study seems to conclusively prove ABC to be a significant cause of bladder cancer if present in drinking water. The results would blow the new standard out of the water (no pun intended). How should he proceed?

6.20.2 Discussion

The information that George received from the international study is truly worthy of his immediate concern. However, it requires a very thorough review by his research staff. After completing LMA's three-year study and convincing the EPA to accept it as a basis for the regulation change, it is difficult to believe that a cancer risk was overlooked. These regulation changes do not occur whimsically. Despite any conclusion in the international study report, there must be some hesitation to accept its conclusion. That international research needs to be carefully examined, and the experimental methods reviewed and evaluated. What were the issues that prompted the study? Cancer risk is not something that one can measure on a short-term basis. The affected population selection can be important. For example, if the population in the test community has an older versus younger age group, this can change the finding. The details of the international tests may not be presented in the report and may not have been taken into consideration.

6.20.3 Recommended Action

Clearly George must take some action. The important thing here is to not panic and to obtain help from other staff members of the association. His primary responsibility is to protect public health. The responsible thing to do is to get the support from LMA's top management and its industry committees necessary for covering the expense of evaluating the international data on ABC's health effects. While this would be upsetting news to the association's members, the caring thing to do would be to not inform them until the international finding is either confirmed or disproved. He needs to respect his friend in Holland and not seek to cover up the international findings, but instead thoroughly evaluate them. In many cases these studies have pros and cons that may have been evaluated to support the advocate's position. This may be true of LMA's own data as well.

George must take the "tip" information he received to as high a level as possible within the association and form an ad hoc team to work the problem. The team decides who and how to approach the international institution that prepared this finding. He must not try to handle this situation alone. He should accept his friend's offer to be a peer reviewer so that he has access to the study details. You may be your team's lead person, but there may very well be another member of LMA that has been in contact with the international group.

George must show respect for the international participants, as well as their data and analysis. Often in a situation like this, the activity can become adversarial. Certainly those 35 percent of facilities in the industry that would be shut down would have hostile feelings due to the other findings. Clearly it could become a charged atmosphere. Therefore, it will be important to keep the issue confined to your ad hoc group until some clarification is established. In a situation like this there are often some established and respected experts who should be included in the study. LMA needs to cover the costs of these experts.

Finally, once George has completed the evaluation, he must deal with the results. If his team finds substantial disagreement with the international findings, he should convey this to his association members and decide who will approach the international association with the findings and how. This is the just and fair thing to do. They will clearly be uncomfortable, but George must respectfully help them, if possible, to accept your evaluation of the data. In the same manner, if LMA's findings are that the international data are correct and its prior research did not discover this aspect, then the association must face the issue of how to control ABC emissions to protect public health.

Either of these situations would be an event of monumental importance, and relationships can become hostile. It is important to keep the pillars of character in mind. Trustworthiness requires that the data are treated truthfully and honestly, only the truth, and the whole truth. Respect requires resisting any hostile feeling and understanding that the investigators on the other side are competent. Responsibility requires handling the issue in a professional manner, keeping informed those who need to be. Justice and fairness require that LMA take an objective view in evaluating the international data and analysis as well as being willing to accept any failings in its own study. Finally, caring means realizing the impacts that LMA's effort will have on the many workers in its industry, and trying to avoid a protracted high-profile release of information in the media.

6.21 Should Consultant Use a Bounty Hunter Provision on Client?

6.21.1 Dilemma

You are an independent consultant performing an environmental audit of PGH's manufacturing plant A. As the consulting auditor, you uncover evidence of an ongoing illegal pollution discharge that has the potential to cause serious harm to public health and safety and the environment. These illegal discharges have never been reported to a regulatory agency. Furthermore, since the discharge has the potential to impact drinking water, it must be reported under a state regulation. You have advised your client by written report to the attention of PGH's plant manager, who refuses to report or disclose the situation as required by law. You are fully aware of the "bounty hunter" provisions under the applicable state regulation, and you are certain that the state would have a strong case against PGH. By reporting this violation to the state agency, and if it is successful in bringing suit against PGH, the court would impose a large fine and you would receive a bounty hunter award as a percentage of that fine. Should you report the incident? How do you proceed?

6.21.2 Discussion

When a consulting environmental professional discovers and reports to the client an illegal discharge that violates emissions regulations, what steps can the consultant take if the client refuses to report the violation? In the interest of protecting public health, it is important that the discharge be reported, that it be discontinued, and that the residue be remediated. As a consultant with proper concern for public health, what can you do in a situation like this? This dilemma involves a situation where the consultant is aware of a law where the person reporting such a violation to the state agency can be rewarded by receiving a percentage of the fine assessed. This is referred to as the bounty hunter provision.

It has been three months since you advised PGH regarding the illegal discharge. You have submitted your detailed report and have been paid for your services. You have not been welcomed in the plant since that time and do not know what, if anything, has happened. You feel a moral responsibility to protect the public from the potential threat from this discharge. You consult an environmental attorney who explains your options, including the bounty hunter provision.

Your attorney tells you that if you take the bounty hunter approach, you will have to change your profession. No one would hire you or any firm for which you work as a consultant in the future. There may be rare occasions when a bounty hunter approach could be taken. But you should know that seldom have these bounty hunter reporters ever benefited in any real way, and their careers are usually permanently damaged. This situation is clearly reportable according to the regulation and your duty to protect public health. It must be reported in order to stop the discharge and initiate the remediation. But reporting the incident to the agency under the bounty hunter provision is not the recommended solution.

To examine this situation further, consider the pillars of character. Trustworthiness requires you to be honest and truthful, to tell the truth and the whole truth. You have the obligation to see that the discharge is reported. Respect requires that you consider all of the stakeholders, the public and the plant employees. To report the condition as a bounty hunter is a selfish act to be used only as a last resort. If the reward is large, this is even more of a selfish act. You do have the responsibility to report the situation in order to protect public health. But convincing the client to self-report is the most responsible action that could be taken.

6.21.3 Recommended Action

As an independent consultant you should show your detailed report to your attorney. He should write a registered letter to the top official at PGH—the resident CEO—and include a copy of the fully documented report of the violation. If you were working in a consulting firm, you would discuss your findings with your consulting firm's supervisor and then elevate the information to the top level of your firm. If your firm has a legal office, it should be consulted. The management team should decide how to proceed together. The person in your firm with the closest high-level relationship with the client firm should arrange a meeting at which the situation is revealed along with the legal requirements. PGH's attorney should be invited to the meeting.

Your attorney should explain the benefit of self-reporting. Any belief on PGH's part that the discharge would not be detected by the agency should be clearly dispelled. The rationale for believing that the discharged material will intrude on drinking water should be presented. The issue of the bounty hunter provision should never be mentioned. Any attorney who is not familiar with this aspect is unlikely to be involved. The objective is to have the discharge condition reported and stopped, and remediation started.

6.22 Environmentalist Blockade of Offshore Incineration Ship

6.22.1 Dilemma

You are the environmental officer on the *Volcano*, a ship designed to safely incinerate hospital and infectious waste offshore. The permit requires the

Volcano to follow a prescribed channel and severely limits the hours of operation of the *Volcano* when it enters or leaves the harbor. On one outgoing trip you find the ship surrounded by small watercraft operated by members of Greensleeves, a powerful and savvy environmental organization. Over the loudspeaker, Greensleeves has declared its intention to blockade your progress, at any cost. It will be difficult to maneuver between them without angering the protesters. Going around them requires leaving the prescribed channel and possibly risking accidental grounding of the *Volcano*. To stay put means exceeding the operating time limits specified in the ship's permit. The captain wants to go full steam ahead, in spite of Greensleeves. How do you proceed?

6.22.2 Discussion

Occasionally there is an adversarial relationship between the environmental professional and the environmentalist. While the objectives of the two activities are often similar, their methods are not always compatible. Here is a case where the two factions clash.

The Greensleeves blockade is an egregious action to demonstrate its protest of offshore waste incineration. This type of incident is one that an environmental professional might face in a rare situation. The ship's captain is in full charge. You can only advise him as to the rules and regulations that involve the timing and efficiency of incineration. But as a person involved with the ship's operation you have the right to state your opinion. The ship should avoid running down the small boats. But if they steer directly into your path this might be unavoidable.

6.22.3 Recommended Action

Therefore, you must call or radio the port authority and request patrol boat and helicopter assistance, as well as advice regarding how to interact with the small boats until the authorities arrive. You suspect that the media has been alerted by Greensleeves, who wants to make this into a serious demonstration against offshore incineration. Your ship must keep making slow headway so as to not drift out of the channel. The captain should be encouraged not to go full steam ahead to prevent a serious collision. If available, the ship's company deck hands should have life jackets and buoys to throw overboard in the event that any of the small boaters should be thrown into the water. The demonstrators may be over the top, but you must prevent them from drowning if possible. Since your operation is legal, the authorities should take the necessary action to break up this protest action and bring disciplinary action against the protestors.

6.23 Driving 55 mph Only When the Cop Is Around

6.23.1 Dilemma

You are auditing a facility as a private consultant to an important client. In talking to the plant personnel and reviewing production records, you learn that the plant operates substantially beyond its permit limits, causing as much as 50 to 75 percent more pollutant emissions than allowed. You learn that the plant management is intentionally operating in this mode. However, when you were performing your official audit, the plant was running within permitted limits. How do you proceed?

6.23.2 Discussion

There are times when environmental testing consultants are taken advantage of by plant operators. The tests are run at allowable plant maximum operating conditions, but at other times the process operates beyond the permit limits. Is it the consultant's responsibly to make this determination and report it?

This is as typical an ethical situation as one can find in this profession. You are aware that air pollution does affect public health; however, one does not see people dropping in the streets grasping at their throats as a result of it. As a private consultant you have no legal policing responsibilities. On the other hand, the plant management is cheating by exceeding its allowable production rate. If every plant in town exceeded its allowable emissions by 50 to 75 percent, there would be a significant impact on public health and welfare. Since the client's management is aware of this arrangement, if you called the violation to its attention, the response would likely be, "So what; when is the last time you drove at 55 mph on the freeway? Let them catch us." It would probably be suggested that your audit report not mention the high production rates, as the client "does not want a paper trail."

Some consultants might comply with the client's wishes, feeling that it is the plant manager's problem if the agency finds out, and that the consultant has discharged his or her responsibility by ensuring that the manager knows the extent of the violation. Here the issue of responsibility and accountability is involved. The adage "all that is necessary for evil to triumph is for good people to do nothing" applies here. Some action must be taken to correct the situation. Also, there are issues of trustworthiness involved, i.e., honesty, integrity, and promise keeping. In performing an audit for the client corporation, the consultant pledges to identify deficiencies so that they can be corrected. The corporate policy that initiated the auditing procedure was endorsed by the corporate board of directors (if not initiated by them) as a means to ensure compliance. The failure to report the condition is a violation of trust. When the agency eventually learns of this practice and takes legal action, the plant manager may lose his job, but the consultant will be called into the spotlight for not discovering the problem. So, what should be done?

6.23.3 Recommended Action

First, recognize the situation for the moral dilemma that it is and do not attempt to resolve the issue alone. Make no immediate commitments to the client company. Bring the situation to your consulting firm's management. If you are an independent consultant, discuss the situation with a colleague.

Then, use the virtue of caring to objectively examine the alternative solutions and seek to minimize repercussions. Have a senior consultant explain to the plant manager the serious implications that will be involved when the agency learns of this illegal practice, including heavy fines under the Clean Air Act and even possible plant closure. Explain the Title V permitting procedure, which might require continuous emissions monitoring. Discuss the possible control options that would allow the plant to continue with these higher production rates without increasing emissions. Include this information in the audit report. As appropriate, send copies of the report to the client's corporate officials and corporate counsel. The wording of the report should be carefully considered to relate facts without insinuation. Within the virtue of caring, it would not be unethical to accept some editing suggestions from the client's legal staff as long as a clear recognition of the situation is not lost.

6.24 Biased Agency Subordinate Stonewalls a Project

6.24.1 Dilemma

You are a senior manager in your state's environmental agency and have just been assigned to coordinate an ongoing case. The case is before the hearing board, and it concerns an incinerator that needs, among other permits, a NPDES permit to discharge wastewater to a large lake. You are reviewing the Fish and Wildlife testimony, which states that irrevocable damage to the native fish population will occur. You doubt the conclusion and contact the person giving the testimony. The Fish and Wildlife expert says: "Well, using that incinerator is a bad idea. Everyone should recycle. The fish argument is the only way you and I can stop it." Your expert will be on the witness stand in a week and the written testimony has already been submitted. How do you proceed? Consider your responsibility to your agency, yourself, the applicant, and the general public health.

6.24.2 Discussion

Occasionally an environmental professional meets an agency employee who believes that it is his or her mission to reduce pollution even if it means going beyond the restrictions that are in the current regulations. The attitude is that you cannot be too safe. In this dilemma we meet such a regulator.

With the information provided in this dilemma, it must be assumed that the agency has no problem with the air emissions and the environmental impacts except for the fish damage argument. Apparently, this staff person has invented an impact, "irrevocable damage" to the fish, which you question. His explanation has nothing to do with the issue. He is against any waste incineration. As a trustworthy, respectful, and responsible public servant, you should not allow the whim of a staffer to stop a project. You may want to show some caring for the staffer and understand that he has only the best of intentions. However, this project is proposed for the benefit of the citizens, and there is a formal standard for having the project approved under the National Pollutant Discharge Elimination System (NPDES) standard.

6.24.3 Recommended Action

Unless the staffer can prove to you and another staff person specific evidence of damage, you should have the written testimony withdrawn and have his scheduled testimony cancelled. How you deal with him depends on your experience with his work and general attitude toward the agency. He has made a serious misjudgment and should be well reminded of it. Agency planners and enforcement personnel can become carried away with the importance of their jobs. It is recommended that an occasional ethics workshop be convened to discuss such issues and thereby avoid embarrassing situations like this one.

6.25 Permitting Engineer Softens NSR Rules to Help Local Industries

6.25.1 Dilemma

You are the air pollution control officer (APCO) of a large county agency. You are aware that current New Source Review (NSR) rules are restricting the economic growth of rural areas located downwind of your city. Because of transport of the pollution from the industries in your large city, that rural region is classed as "severe nonattainment." If no industry existed in that region, it is likely that it would still be classified severe nonattainment.

You are alerted that a permitting engineer in your agency has been using a relaxed interpretation of NSR rules. Companies have been allowed to buy other companies and merge their potentials to emit, and even allowed to physically merge the companies in one contiguous area. When you confront the agency's engineer, you learn that she believes that the area has suffered disproportionately by the latest slowdown, unemployment is high, and this minor concession is a way to help people. You assure yourself that this person has not benefited materially in any way. You are also aware that the EPA regional auditors will conduct their annual audits in two months. How do you proceed?

6.25.2 Discussion

This dilemma is especially interesting because is not a hypothetical situation; it really happened. In contrast to the previous dilemma, the agency person arbitrarily raises the allowable emissions to assist industry in her town that is experiencing poor economic conditions.

There are times when a regulation can unintentionally and unfairly impact a local region. Your permitting engineer is a caring person. She is single and a graduate of a prestigious state school. You recall her great attitude when you interviewed her for the agency job. Her act is not selfish, but it is inappropriate.

You are keenly aware of this complex regulatory problem in this downwind area. While you understand the basis for the NSR regulation, you suspect that the EPA regulators, when they defined the NSR rule, did not consider this possible exception. This woman's compassion has now forced your consideration.

6.25.3 Recommended Action

First, you must sit down with the agency's engineer, and possibly the entire agency staff, and explain the situation. You should have already made up your mind that you will elevate this situation to the state agency or regional EPA (or equivalent) office. If there are Title V concerns, the EPA must be contacted. It would be appropriate to obtain the support of city officials and even a local state assembly representative or a federal congressperson if you can. You explain this inconsistency between two federal regulations to your agency staff at your meeting and allow them to offer any recommendations that they may have. Allow for constructive discussion. In some respects this is not truly an ethical dilemma, but doing the right thing may be difficult and should not be handled superficially or deceptively.

Glossary

A&WMA: Air and Waste Management Association.

AACSB: Association to Advance Collegiate Schools of Business.

ABET: Academic Accreditation Board for Engineering and Technology.

Anthropocentrism: Anthropocentrism assumes that the environment exists for serving humans, a human-centric view.

API: American Petroleum Institute.

ARCS: Alliance for Research in Corporate Sustainability.

Bait and switch: An expert is presented in the proposal and less qualified alternatives are used in the implementation of the project.

BEE: Business Ethics Everywhere.

Benefits monetization process: Estimating an economic value from the benefits of an activity.

Benzene emission credits: Reducing benzene emissions below mandated allowable levels for selling the generated credits in the market.

Biocentrism: Places the biological world at the center of the planet and focuses on the intrinsic value of life.

Biodiversity/biological diversity: Describe the variety of life on earth or the degree of variation of life-forms within a given species.

Biophysical view: This perspective is based on the view that living systems are constantly exchanging energy with information at multiple levels.

BOP: Base of the pyramid.

Bosons: Recently discovered subatomic particles.

Brahmo Viharas: Translated as the resting place of the divine spirit in Buddhism.

CAFE: Corporate Average Fuel Economy.

Carbon share: The amount of carbon associated with meeting any human need for products or services.

CBD: Convention on Biological Diversity.

CEQA: California Environmental Quality Act.

CERCLA: Comprehensive Environmental Response, Compensation, and Liability Act.

Cheating: Achieving a goal by misleading, e.g., green washing, lying.

CITES: Convention on International Trade in Endangered Species of Wild Fauna and Flora.

Classification of goods in economics: Goods are rival (consumption of one limits use by another), nonrival (enjoyed by many at a time), excludable (only the buyer enjoys), nonexcludable (access open to all), private (can be assigned property rights), or public (no property rights, typically nonrival and nonexcludable).

CLP holdings: China Light and Power Holdings Company.
CMS: Convention on the Conservation of Migratory Species of Wild Animals.
Collusion: Stealthy or unlawful partnership to cheat others.
Command and control: Prescribed directive, usually flowing from the top down from positions of authority.
Commodity: A marketable item that satisfies a need or a want; most environmental goods are public and not marketable as a commodity, e.g., the breathtaking view of Mount Everest.
Contingent valuation method: A way to value nonmarket goods, e.g., environmental goods using surveys.
Controls (pollution): Equipment or operating practices to reduce pollution.
Core ethical values: Josephson defines ten core values—honesty, integrity, respect, caring, fairness, promise keeping, the pursuit of excellence, civic duty, accountability, and loyalty.
Corporate image: Perception of the business by stakeholders.
Corporate manslaughter: A homicide committed by a company.
Cost–benefit analysis: Systematic estimate of equivalent money value of costs and benefits of an action.
CSI: Creating social impact.
CSR: Corporate social responsibility.
CSV: Creating shared value.
Culpability: Blame that is attributable.
***Dama*:** Counsels control of passions.
***Dana*:** Enthuses one to give others their due share.
***Daya*:** Urges one to treat all of God's creatures with kindness and compassion.
***Dharma*:** Cosmic contract of individual rights and duties.
DJSI: Dow Jones Sustainability Index.
Downside risk: Potential for loss that could be the result of an event or action.
Ecological economics: Addresses relationships between ecosystems and economics and deals with preservation of natural capital.
Ecosystem: Community of living organisms and nonliving components of nature that are interdependent for sustenance.
Ecosystem service: Resources and processes supplied by natural ecosystems.
EIR: Environmental impact report.
EIS: Environmental impact statement.
Emission hot spots: Areas of localized high emission levels.
Emissions fees: A fee typically charged by a government per unit of emission.
Enlightened self-interest: Realization that there is value in other's interests, because it is essential to also serve one's long-term overall self-interest.
Environmental advocacy: Presenting information on nature and environmental issues to encourage environmentally sensitive attitudes.
Environmental economics: Deals with economic effects of environmental policies.

Environmental ethics: Per *Stanford Encyclopedia*: "Environmental ethics is the discipline in philosophy that studies the moral relationship of human beings to, and also the value and moral status of, the environment and its nonhuman contents" (accessed December 2012, http://plato.stanford.edu/entries/ethics-environmental/).

Environmental ethics culture: Characteristics and behavior of an organization in the context of environmental ethics.

Environmental professional: One whose primary responsibility is to protect public health and the environment.

Environmental space: Various aspects within the environmental area.

EPA: Environmental Protection Agency.

Equator Principles: Voluntary standards commitment by project lenders to manage social and environmental risk.

Equi-marginal principle: Marginal cost of control across polluters to achieve emission reduction is equated at the lowest possible cost.

ERC: Emission reduction credits.

ESG factors: Environmental, social, and governance factors.

Ethical behavior: Implicit in it are moral duties and virtues defined by core ethical values that form the philosophical basis for ethical judgments.

Ethical conduct: When codified it becomes a guideline for appropriate individual and collective behavior, the minimum of which is moral duties.

Ethical decision making: Process of competently identifying issues and evaluating consequences, seeking different opinions, deciding what is right with self-confidence, and exercising willingness to make decisions when the issue has no clear solution.

Ethical dilemma: Involves a complex situation where the choice of an alternative produces winners and losers among the stakeholders and leads to a conflict in decision makers' minds among ethical imperatives.

Ethical principles: Rules of conduct derived from values that pertain to what is right or wrong and motivated by a sense of moral duty.

Ethical relativism: The use of ethics that favors the considerations of the nonjudgmental agent with no ethical evaluation of actions.

Ethical values: Values that directly relate to beliefs concerning what is right and proper or which are motivated by a sense of moral duty.

Ethically neutral: Neither ethical nor unethical.

Excludability: Ability to keep something away from people that do not pay for using it.

Exposure, acute and chronic: Coming into contact with a substance; acute as in short-time but high-level exposure or chronic or long-term because of lower-level but repeated exposure.

FTSE4Good Index: An index to measure the company performance relative to a set corporate responsibility standard.

GHG: Greenhouse gas that contributes to the phenomenon of global warming, e.g., carbon dioxide, methane.

Green washing: Intersection of poor achievements coupled with false positive communication about environmental performance.
GRI: Global Reporting Initiative.
Group think: Where the desire for harmony in a decision-making group overrides a realistic appraisal of alternatives.
Hadith and sunnah: Prophet Muhammad was responsible for spreading the teachings of Islam and leading a life of example; his words are known as *hadith*, and his actions are called the *sunnah*.
Hazardous waste: Discarded hazardous substances become hazardous waste; also material left after products are made in a manufacturing process.
HSWA: Hazardous and Solid Waste Amendments.
IFC: International Finance Corporation.
Initial assignment of property rights: Prior to the emergence of the optimal efficient solution, there is a need to assign an initial "right to emit," e.g., a level of emission rate.
IPCC: Intergovernmental Panel on Climate Change.
IPO: Initial public offering.
Is ethics—descriptive ethics: Only describes operational standards of behavior—how an individual or group actually behaves—without reference to judgments of right and wrong.
IUCN: International Union for Conservation of Nature.
Laissez-faire: Autonomous economic order with no government intervention.
LEED: Leadership in Energy and Environmental Design.
License to operate: Permission from community to operate the facility within.
Lower eco-strata: Lower end of the spectrum of socioeconomic status defined by family income, parental education level, parental occupation, and social status in the community.
Machiavellianism: The willingness to do whatever it takes to achieve a goal.
Malthusian: Perspective of economist Simon Malthus.
Many to many: Many people to many people.
Market efficiency: A market is efficient if the price of the goods transacted is an unbiased estimate of the true value of the goods exchanged, where errors are random.
Market externality: Where the cost and benefits do not transmit in the exchange through the price associated with the transaction.
MNC: Multinational corporation.
Moral duty: Implies conduct that is right, ethical, or proper.
Moral obligation: *See* moral duty.
Moral virtue: Goes beyond moral duty; is not mandatory but something to be aspired to, e.g., generosity or bravery.
Multiverse: Multiple possible universes.
Natural attenuation: Reduction in mass or concentration of a compound in the environment due to naturally occurring physical, chemical, and biological processes.

Negative externality: Where the social cost of market activity is not covered by the pricing in the economic exchange.

NEPA: National Environmental Policy Act—an act to establish a national policy for the environment (1969).

Net Impact: A nonprofit inspiring a new generation to use their careers to tackle the world's toughest social and environmental problems.

Next economy: The next phase of the economy driven by the new business ambiance to meet the new social contract.

NGO: Nongovernmental organization.

NIMBY: Not in My Back Yard.

NOAA: National Oceanic and Atmospheric Administration.

Nonethical values: Values that are ethically neutral.

Norms of behavior: *See* ought ethics.

NSF: National Science Foundation; a U.S. agency that funds fundamental scientific research.

OECD: The Organization for Economic Cooperation and Development.

ONE: Organization and the Natural Environment.

OSHA: Occupational Safety and Health Administration, a U.S. agency.

Ought ethics—prescriptive ethics: Involves revelation of and commitment to principles that establish norms of behavior applicable to every person; describes how a person should behave based on specific values and principles that define what is right and proper.

Permit transaction: Documentation administration work related to permit.

Phase I property assessment: Involves an exhaustive background paperwork examination and an on-site inspection to assess need and scope of a phase II assessment.

Pigovian tax: A tax on negative externalities a market activity generates.

Positive externality: Where there are public benefits from a market activity but inadequate margins cause undersupply of product and government subsidies are needed to avoid that.

Power of largess: First get rich, then get generous.

Principled reasoning: Justification based on the ethical principles.

Probability/consequence screening matrix: A matrix of probability and consequence of events on two axes for use as a screening tool for prioritizing risks.

R&D: Research and development.

Ramsar: Convention on Wetlands of International Importance.

RCRA: Resource Conservation and Recovery Act.

REACH: Registration, Evaluation, Authorization, and Restriction of Chemicals.

Regulatory agency: The government authority responsible for environmental regulation.

Religious values: Those values specified by a godly figure and subject to the interpretation of the clergy, or by philosophers of particular cultural groups.

Reporting facility: An operating facility subject to reporting to a regulatory agency.

Resource allocation: Assigning parts of the total resources on some set basis.

Resource economics: Economics of production and use of natural resources.

Risk assessment: The identification of risks followed by their evaluation or ranking.

Risk management: The assessment of significant risks and the implementation of suitable risk mitigation responses.

ROD: Record of decision.

SAM: Sustainability asset management—an investment adviser firm.

Sarbanes–Oxley Act: Sarbanes–Oxley requirements.

SEA: Social and environmental impact assessment.

SEC: Securities and Exchange Commission.

Self-interest: Inherently inclined to promote own benefit; each out for oneself.

Shareholder primacy: As opposed to all major stakeholders' primacy, only the shareholder's interest is given consideration.

Shareholder value devastation: Shareholder wealth is destroyed or very significantly reduced.

Shareholder wealth: Value of the shareholder's stock ownership in the company.

Share price: The price of one share of company's saleable common stocks.

Shariah: A set of natural laws created by God (Islam).

Simon–Ehrlich wager: Economists Julian Simon and Paul Ehrlich entered a bet in 1980 on resource scarcity over the decade leading up to 1990.

Six pillars of character: Josephson defines moral duties and virtues that flow from the six pillars of character: trustworthiness, responsibility, respect, justice and fairness, caring, and civic virtue and citizenship.

Social contract: An explicit, legal, and implied contract between the corporation and the organizations on the one side and the society or community of stakeholders it works within and is a part of on the other.

Social discount rate: The mutually agreed rate used in policy decisions to discount social costs and benefits in the future; cost–benefit analysis frameworks use this discount rate to compare costs and benefits received at different points in time.

Society: An association of individuals or an organized group brought together because of common interests (living in the same demographic region), beliefs (religion, politics, or cause), or profession (company or field of work).

Society's willingness to pay: The portion of society's resources it is willing to allocate or pay.

Solovian: Perspective of economist Robert Solow.

Spatial distribution: Distribution of a trait over space, e.g., contaminant or species.

Species diversity: Measure accounts for both species diversity and profusion.

Species richness: Count of number of different species.

Stakeholder: Entity or group that has an interest in and could be impacted by or could impact an action.

Stakeholder primacy: As opposed to shareholder primacy, all major stakeholders' interests receive just and equivalent primary consideration.

Superfund: Common name for the Comprehensive Environmental Response, Compensation, and Liability Act of 1980 (CERCLA), the U.S. environmental program established to clean up abandoned hazardous waste sites, the law created a tax on chemical and petroleum industries.

Sustainable development: Meets the needs of the present without compromising the ability of future generations to meet their needs.

Synchronous interactive connectivity: Simultaneous linked communication through digital media.

Tanakh: The Hebrew Bible or the Old Testament.

TEEB: The Economics of Ecosystems and Biodiversity.

Torah: Means teaching or the law—the collection of the first five books of the Hebrew bible.

Toxic tort: A lawsuit that involves impact of a toxic substance.

Tragedy of the commons: Tragic depletion of shared natural resources when individuals ignore common good and act entirely in their self-interest.

Transaction costs: Cost incurred in a market exchange, e.g., legal enforcement of a regulatory mandate.

TRI: Toxic Release Inventory.

TSCA: Toxic Substances Control Act.

UNFCC: United Nations Framework of Climate Change.

Upside risk: Potential for loss that could be the result of an event or action.

VOSL: Value of statistical life.

VOSL years extended method: Estimates the period by which the value of statistical life-years is extended because of a protective action or regulation.

WBCSD: World Business Council for Sustainable Development.

Whistleblower: Someone who informs a supervisor, a regulator, or the public about a suspected unethical or illegal activity.

World becoming flat: A term popularized by Tom Friedman's book *The World Is Flat*, meaning the globalization of all businesses.

Appendix

Air and Waste Management Association's Code of Ethics

(Policy Manual: Section 8.7)

AIR & WASTE MANAGEMENT
ASSOCIATION
SINCE 1907

PREAMBLE: In the pursuit of their profession, environmental professionals must use their skills and knowledge to enhance human health and welfare and environmental quality for all. Environmental professionals must conduct themselves in an honorable and ethical manner so as to merit confidence and respect, as well as to maintain the dignity of the profession. This code is to guide the environmental professional in the balanced discharge of his or her responsibilities to society, employers, clients, coworkers, subordinates, professional colleagues, and themselves.

PLEDGE: As an environmental professional, I shall regard my responsibility to society as paramount and shall endeavor to:

1. Direct my professional skills toward conscientiously chosen ends I deem to be of positive value to humanity and the environment; decline to use those skills for purposes I consider to conflict with my moral values.
2. Inform myself and others, as appropriate, of the public health and environmental consequences, direct and indirect, immediate and remote, of projects in which I am involved, consistent with both standards of practice in industry and government, as well as laws and regulations that currently exist.
3. Comply with all applicable statutes, regulations, and standards.
4. Hold paramount the health, safety, and welfare of the public, speaking out against abuses of the public interest that I may encounter in

my professional activities, as deemed appropriate per professional standards and existing laws and regulations.

5. Inform the public about technological developments, the alternatives they make feasible, and possible associated problems wherever known.
6. Keep my professional skills up to date and endeavor to be aware of current events, as well as environmental and societal issues pertinent to my work.
7. Exercise honesty, objectivity, and diligence in the performance of all my professional duties and responsibilities.
8. Accurately describe my qualifications for proposed projects or assignments.
9. Act as a faithful agent or trustee in business or professional matters, provided such actions conform to other parts of this code.
10. Keep information on the business affairs or technical processes of an employer or client in confidence while employed and later, as required by contract or applicable laws, until such information is properly released and provided such confidentiality conforms with legal requirements and other parts of this code.
11. Avoid conflicts of interest and disclose those known that cannot be avoided.
12. Seek, accept, and offer honest professional criticism, properly credit others for their contributions, and never claim credit for work I have not done.
13. Treat coworkers, colleagues, and associates with respect and respect their privacy.
14. Encourage the professional growth of colleagues, coworkers, and subordinates.
15. Report, publish, and disseminate information freely, subject to legal and reasonable proprietary or privacy restraints, provided such restraints conform to other parts of this code and do not unduly impact public health, safety, and welfare.
16. Promote health and safety in all work situations.
17. Encourage and support adherence to this code, never giving directions that could cause others to compromise their professional responsibilities.

Index

A

Accreditation Board for Engineering and Technology (ABET), 115
Acid rain, 47
Advocacy, environmental. *See* Environmental advocacy
Agnosticism, 31
Air and Waste Management Association (A&WMA), 105
Alberta Specified Gas Emitters Regulation, 56
American dream, 70
Anthropocentrism, 38–39
Aquinas, Thomas, 38
Arrhenius, Svante, 55
Atheism, 31
Augustine, Norman R., 103

B

Basel Convention, 54
Biocentrism, 39–40
Biodiversity, 60–61, 62
Bioethics, 60–61
BP Gulf oil spill, 8–9, 9
Brown, Pat, 62
Brundtland Commission, 46, 83
Buddhism, 28–29
Buffett, Warren, 71, 94, 95
Bush, George W., 7
Business Ethics Everywhere (BEE), 119
Business school, ethics training, 116–119

C

California Environmental Quality Act (CEQA), 63
Callahan, David, 5, 6
Capitalism, 70
Carbon share, 55–56
Caring, 17, 22
Carson, Rachel, 53, 84
Case examples, 148–150
  agency alters rules to help local industry, 231–232
  bad mouthing competitors, 185
  bait and switch, 175–176
  bias, 230–231
  billing issues, 210–212
  bounty hunter provisions, 226–227
  bribes, 148–150, 157–159
  client-attorney privilege, 200–203
  compliance issues, 144–146
  confidentiality, 168–169, 203–206
  contract assignments to friend, 167–168
  data suppression, 182–185, 223–225
  dishonest use of company resources, 214–216
  employee writing book, 152–154
  enforcement issues, 220–221
  environmentalists, 227–228
  errors in data, 171–172
  expert witness, 138–139
  extortion, 206–208
  falsifying data, 142–143, 164–165, 172–274
  hazardous wastes, 161–162, 170–171, 216–218
  hiding finings, 129–131
  media issues, 222–223
  modeling issues, 174–175
  money demands, 135–137
  moonlighting employee, 146–148
  not reporting negative findings, 179–182, 197–200
  passing the buck, 177–179
  permit issues, 131–133, 155–157, 163–164, 208–210, 229–230
  political issues, 159–161
  pricing issues, 133–135
  recruiter encourages you to bring customers/staff, 187–191
  reporting on a competitor, 195–197
  risk communication, 218–219
  sexual harassment, 140–141, 141–142

silent partner, 193–195
subterfuge, 165–166
whistleblowers, 151–152
Cato Institute, 52
Character, pillars of, 17, 136; *see also specific character traits*
Cheating culture, 5, 6, 75
Chick-fil-A, 81
Christianity, 24–25, 29, 38
Citizenship, 17, 22, 83
Civic virtue, 17, 22, 132
Clean Air Act, 56
Client–attorney privilege, 200–203
Climate change
carbon share; *see* Carbon share
consensus regarding, 55
fossil fuels, role of, 57
government interventions regarding, 56–57
United Nations Framework Convention on Climate Change (UNFCCC), 55
Clinton Global Initiative, 71
Clinton, Bill, 71
Coase, Ronald, 45, 47
Collusion, 9
Committee of Sponsoring Organizations of the Treadway Commission, 50
Common-law principle, 192
Competition, 146
Compliance, 144, 145, 163
Compliance and Ethics Leadership Council, 84
Compliance codes, 20
Compliance, ethical, 94
Compliance, legal, 94
Comprehensive Environmental Response, Compensation, and Liability Act (CERCLA), 59, 88
Confidentiality, 129, 168–169, 182, 203–206
Convention on Biological Diversity (CBD), 61
Convention on International Trade in Endangered Species of Wild Fauna and Flora (CITES), 61
Convention on the Conservation of Migratory Species of Wild Animals, 61
Convention on Wetlands of International Importance, 61
Cooper, Cynthia, 7, 8
Corporate Average Fuel Economy (CAFÉ) standard, 52
Corporate governance, 84
defining, 92
environmental responsibilities, 86–87
fiduciary duties, 85, 86
principles of, 92–93
Corporate social responsibility (CSR)
debate regarding, 69
enlightened self-interest, as policy of, 69–70
evolution of, 67–68
public relations regarding, 72
reporting, 81, 8890
stakeholders, role of, 69; *see also* Stakeholders
strategic, 70–72
Corporate stewardship, 54–55
Cost–benefit analysis, 47–49
Creating shared value (CSV), 75
Creating social impact (CSI), 73–75, 81
Cultural values, 13–14
Culture
American, 13
cheating, of, 5, 6, 75
ethical, 9–192

D

Defense, Department of, 9
Delta-Mendota Canal, 63
Dow Jones Sustainability Index, 75, 87
Dow Jones Sustainability Index93
Druyvesteyn, Kent, 119

E

Economics, environmental. *See* Environmental economics
Ecosystem
protecting, 3
Ehrlich, Paul, 52
Emission reduction credits (ERCs), 47

Energy Information Administration, 57
Enron
  attorneys for, 9
  blame game, 9–10
  greed, role of, 9
  whistleblowers, 7, 8, 9
Environmental advocacy, 53–54
Environmental economics
  commons, tragedy of, 45
  cost–benefit analysis; *see* Cost-benefit analysis
  defining, 42
  economic instruments, 46–47
  goods, classification of, 44–45
  human capital approach, 49
  market externality, 43–44, 45
  markets, myth of, 42–43
  Mathusian and Solovian hybrid, 51–52, 68
  private market, 42
  property rights, 45–46
  regulatory policy; *see* Regulatory policies, environmental
  resource prioritization, 50–51
  risk management; *see* Risk management
  universal market, 42
Environmental ethics training
  business school, ethics training, 116–119
  code of, 105, 106–107
  evolution of, 105
  facilitators, 109, 110
  need for, 103–104
  overview, 103
  requirement of, at graduate level, 115
  structure of, 108–110
  workshops on, 107–108
Environmental impact report (EIR), 61–62
Environmental impact statement (EIS), 39, 41
Environmental justice, 43
Environmental professional (EP) firms, contracting with, 20–21
Environmental Protection Act, 41
Environmental Protection Agency (EPA), 5, 9, 56, 133, 172, 173, 174, 223
Environmental regulatory policies. *See* Regulatory policies, environmental
Environmental studies, 115–116
Environmental, social, and governance (ESG) metrics, 88
Environmental, social, and governance (ESG) performance, 67
Environmentalists, 227–228
Equator Principles, 86
Equity, 95
Ethical behavior, 3, 31–32
  decisions involved in, 10
  nature of people, 4
  reasons to act as such, 14, 15–16
  self-interest, *versus,* 14
  training in, 4, 6
  unethical culture, *versus,* 5–6
Ethical decision making, 12–13, 76–78, 79–81
Ethical dilemmas, recognizing, 16–17, 104, 110–111, 114, 130
Ethical principles for environmental professionals, 82–83
Ethical values, 12, 14, 95
  environmental professionals, for, 82
Ethically neutral values, 12
Ethics. *See also* Ethical behavior
  code of, 79, 92, 211
  defining, 11, 96
  human-centered, 41
  monitoring of, 95
  morals, role of; *see* Morals
Ethics Resource Center, 96
Externality, market, 43–44, 45
Externality, negative, 44
Extortion, 206, 207–208

F

Fairness, 17, 21–22, 83, 95, 199
Federal Bureau of Investigations (FBI), 7
Federal Sentencing Guidelines, 5
Four-way test, 79
FTSE4Good Indexes, 93

G

Gandhi, Mahatma, 9, 28, 76, 120

Gates, Bill, 71
Gates, Melinda, 71
Global stewardship, 92, 97
Global warming. *See* Climate change
Global Warming Solutions Act of 2006, California, 56
Goldman Sachs, SUSTAIN, 87
Goupthink, 78
Great Depression, 84
Green washing, 72
Greenhouse gases (GHGs), 55
Greenspan, Alan, 3–4
Gulf oil spill. *See* BP Gulf oil spill

H

Hardin, Garrett, 45
Hayward, Tony, 8–9
Hazardous and Solid Waste Amendments (HSWA), 59
Hazardous wastes, 58, 59–60, 161, 162, 169–170, 216–218
Hinduism, 26–28, 38, 39
Honesty, 14
  importance, 18
  trustworthiness, relationship to, 18
Hydropolitics, 62

I

Initial public offerings (IPOs), 85
Innovation, 52–53
Integrity, 14, 18, 94
International Finance Corporation (IFC), 86
International Integrated Reporting Community (IIRC), 91
International Union for Conservation of Nature (IUCN), 60, 62
Islam, 29–31
ISO 31000, 50

J

Josephson, Michael, 12–13, 19
Judaism, 24–25, 25–26, 29, 38
Justice, 17, 21–22, 32, 199

K

King, Jr., Martin Luther, 8
Kyoto Protocol, 54, 55

L

Leadership in Energy and Environmental Design (LEED), 52
Leadership, ethical, 119–121
Lockheed Martin, 103
Love Canal, 54
Loyalty, 191
  employer, to, 19
  family, to, 15, 19
  prioritizing, 19
Lying, concealing, 18

M

Madoff, Bernard, 8, 10
Malthus, Thomas, 52, 68
Media/journalism, 222–223
Millennium Ecosystem Assessment, 60
Misleading facts, 18
Molina, Mario, 55
Montreal Protocol, 54
Moral principles, 18
Morality, 11
Morals, 11–12
Moussaoui, Zacharias, 7
Mueller, Robert, 7

N

National Environmental Policy Act (NEPA), 62–63
Nature, preservation of, 40–41
Neutral values, ethically, 12
New social contract, 84
NIMBY (Not in My Back Yard) syndrome, 60
Noncompliance, 76
Nongovernmental organization (NGO), 39

O

Office of Management and Budget, 48
Operation Ill Wind, 5
Organization for Economic Cooperation and Development (OECD), 92

P

Paine, Lynn Sharp, 6
Philanthropy, 71
Philosophical values, 13, 23
Plato, 31
PLUS method, 96–97
Pollution Prevention Pays (3P) program, 93
Ponzi schemes, 10
Poverty, 68, 72
Purpose, triple bottom-line, added to, 3–4

R

Ramsar Convention, 61
Reciprocity, 32
Registration, **E**valuation, **A**uthorization, and Restriction of **Ch**emical substances (REACH), 59, 60
Regulatory policies, environmental, 46–47
Office of Management and Budget, analysis of, 48
Reliability, 83
Religious values, 13–14, 23, 24. *See also specific religions*
Resource Conservation and Recovery Act (RCRA), 59
Respect, 17, 19–20
Responsibility, 17, 20–21
Responsiveness, 83
Risk communication, 218–219
Risk management, 50–51
Ross School of Business, 72
Rotary Club, 96
Rotary International, 79
Rowley, Coleen, 7

S

Salomon Brothers, 94, 95
Sarbanes–Oxley requirements, 50, 84
Sarni, Vin, 119
Scott, Lee, 93
Securities Exchange Commission (SEC), 56
Self-interest, 3, 14
Self-interest, enlightened, 31, 32, 69–70
Self-preservation instinct, 3
Self-sacrifice, 103
September 11 attacks, 7
Sexual harassment, 140–141, 141–142
Shareholders, social responsibility, *versus*, 8–9
Silent Spring, 53
Simon, Julian, 52
Simon–Ehrlich wager, 51
Singer, Peter, 31, 39
Slovanian innovation, 52–53, 68
Social contract, new, 85
Socrates, 79
Solow, Robert, 46
Stakeholders, 3
- diversity of, 95
- synchronous interactive connectivity, 67
- whistleblowers, of, 7

Stewart, Potter, 96
Subterfuge, 165–166
Superfund, 22, 54, 59
Sustainability, 3, 46
- case for, 96
- development, 53–54
- Dow Jones Sustainability Index; *see* Dow Jones Sustainability Index
- metrics, 88–89, 89–90
- reporting, 87
- water stress issues; *see* Water stress

T

Talmud, 26
Tobacco companies
- indictment of, 10

Torah, 25
Toxic Release Inventory (TRI), 59
Toxic Substances Control Act (TSCA), 59

Transparency, 82
Triple bottom-line, 3
Trustworthiness, 17–19, 197

U

Unethical behavior. *See also* case examples
  common situations in corporations, 75, 76
  media coverage of, 16
United Nations Framework Convention on Climate Change (UNFCCC), 55
United Way, 71

V

Valuation, 47–49
Values
  checklist for, 114–115
  cultural; *see* Cultural values
  ethical; *see* Ethical values
  philosophical; *see* Philosophical values
  religious; *see* Religious values

W

Walmart, 93
Water stress, 62–63
Watkins, Sherron, 7, 8, 9
Wealth disparity, 68
Whistleblowers, 92, 150–151, 151–152
  difficulties of, 7–8
  Enron case, 7
World Bank, 86, 93
World Business Council for Sustainable Development (WBCSD), 87
Worldcom, 7